THE ART OF BRAINSTORMING

The Practical Guide to Mastering Creative and Design Thinking and Generating Out of the Box Ideas to Solve Personal and Professional Problems

Philip I. Snyder

DREAM BOOKS LLC
ESTD 2020
THE BEAUTY OF YOUR DREAMS

TABLE OF CONTENTS

Introduction 1

    What is Creativity? 2

    Can We Improve Creativity? 5

    Out of the Box Thinking.. 6

    What is Brainstorming? 9

    Why Brainstorming Matters?..................................... 10

    Brainstorming Examples 12

    What Can You Discover in This Book?....................... 18

Chapter 1: Fostering Creativity.................................... 21

    The Essential Ingredients of Creativity..................... 22

    The Essential Thinking Techniques for Creativity.... 24

    Domain Specificity of Creativity 32

Chapter 2: Understanding Brainstorming 37

    History of Brainstorming... 37

    The Principles and Practice of Brainstorming 39

    The Evolution of Brainstorming................................ 44

Chapter 3: Productive Thinking................................... 49

    Productive and Reproductive Thinking...................... 51

    Ideas Having Sex.. 54

    Thinking Like a Genius 58

    Thinking Like a Designer 64

Chapter 4: Brainstorming Tools and Techniques 71

    Shortcomings of Traditional Brainstorming.............. 72

    Questionstorming.. 76

    What-If Questions 81

    Mind Mapping 82

Brainwriting.. 86

Collaborative Brainwriting 90

Reverse Brainstorming.. 94

Starbursting 97

Rolestorming 101

Braindumping.. 104

Brainsteering... 105

Stepladder Brainstorming.. 113

Charrette Procedure 114

Online brainstorming (aka Brain-netting)...................... 115

Chapter 5: Group Brainstorming 123

Is Group Brainstorming Still Relevant?........................... 123

Ten Surefire Ways to Kill a Brainstorming Session 127

The Ingredients of a Winning Brainstorming Session..... 140

Running an Engaging Virtual Brainstorming................... 157

Chapter 6: Individual Brainstorming............................161

Why not Just Group Brainstorming?................................ 163

Individual Brainstorming Techniques............................. 165

Stimulating Creative Thinking 174

The Best of Both Worlds 184

Closing Remarks...195

References..199

SPECIAL BONUS!

Want This Bonus Book for <u>FREE</u>?

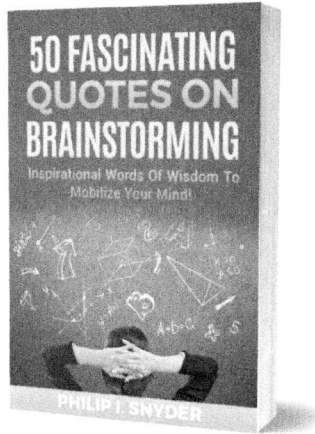

Get FREE, unlimited access to it and ALL of my new books by joining the Fan Base!

Also by Philip I. Snyder

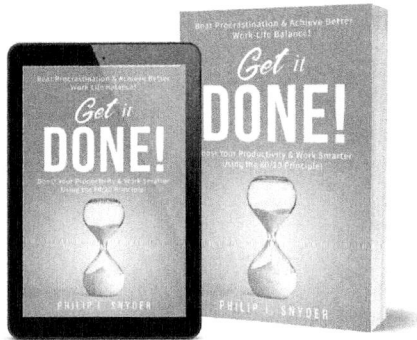

Want to stop procrastinating and start producing? What if you can power yourself through your day and get things done like a pro? Then *Get It Done!* by Philip Snyder is the book for you!

Scan the QR code to get your copy!

COMING SOON!

The practical guide for improving your problem-solving skills in an engaging read, and complementary to *The Art of Brainstorming!*

INTRODUCTION

Have you ever tried to come up with a new cocktail mix idea, a cool theme for a child's birthday party, a new way of rearranging your bedroom furniture, or choosing a farewell gift for a colleague? It is not easy, especially if you try to develop something new or novel as if we tend to fall into the habit of creating variations of the same thing repeatedly.

Imagine pouring water on a flat soil surface. The impact of the water on the soil creates dents and grooves that allow the water to spread out. Imagine doing this several more times, and somehow, the water settles into a specific pathway. It does not create more grooves and dents but focuses on the existing tracks to find its path. That is how most people approach idea generation, falling back on their past experiences to find new solutions. Henry Ford, the legendary American businessman and the founder of Ford Motor, once said, "If I had asked people what they need, they would have said faster horses!" People knew they wanted to travel faster, but most of them (not Henry Ford!) could not think beyond their mental box.

Thinking patterns are like habits. They form over time, and once set, it is hard to change them. While developing specific thinking patterns (i.e., ways to approach problems) saves us a great deal of mental energy in performing our daily tasks, they are usually not helpful to generate novel and creative ideas.

This book is written to answer the following question: "How to generate novel ideas for creative problem-solving?" Perhaps we all agree that fostering a creative mindset is crucial to succeeding in personal and professional life. Take the following example. In 2014, a Michigan-based music band called *Vulfpeck* struggled to get enough resources to fund an upcoming tour. A regular thinking pattern would suggest getting a loan or selling some stuff to make cash. Instead, *Vulfpeck* took a different and brilliant approach; they composed a collection of entirely silent recordings and called it *Sleepify*. Then asked their fans to stream *Sleepify* continuously at night as they sleep! The clever stunt worked, and they earned around 20000 dollars in royalties!

Throughout this book, we will review many more examples and historical stories of how clever ideas could make a difference in our personal and professional lives. The common theme in all of these examples is one word: creativity! So before moving further, let's take a brief look at this key concept. We will discuss creativity and how to foster it in more detail in Chapter 1.

What is Creativity?

Creativity has a loose definition. According to Cambridge Dictionary, creativity is *"the ability to produce or use original and unusual ideas."* At the same time, the Oxford Dictionary defines it as *"the use of the imagination or original ideas, especially in the production of an artistic work."* Indeed, most people think of a person's ability to work on a painting, compose music, write poems, or create a story they talk about creativity. These are all common avenues for creativity, but these are not the only ones. Producing or using original ideas could occur in any field. You can be creative in business, how you teach, how you garden, how you design a house, how you get from point A to point B during a jog, and so on.

According to James E. Kaufman, Professor of Psychology at California State University, there are four major types of creativity: the *Mini-C*, the *Little-C*, the *Pro-C*, and the *Big-C*. Here is a short definition for each type:

- *Mini-C* refers to small personal moments of creativity. For example, when you add a new ingredient to a familiar recipe and make it taste better, or when you try to solve a minor problem through some DIY techniques.

- *Little-C* refers to moments of creativity that other people can recognize. For example, you play the piano, compose a song, or manage to produce crafty materials that others appreciate. This is different from Mini-C, which is creativity that only you recognize.

- *Pro-C* is the creativity of an expert, let's say a scientist, who manages to develop a brand-new way to process garbage and turn it into a useful product. It often takes years of hard work and requires intense focus in answering a pervasive problem.

- Finally, *Big-C* refers to the creativity of a genius, for example, the novels of Mark Twain, Mozart's symphonies, or Isaac Newton's Law of Gravity. Today, some can argue that Apple's iphone, Elon Musk's Tesla, or other disruptive innovations with a lasting impact on our way of life are big-C.

You might wonder why it matters to classify creativity? Well, the classification tells you that everyone can be creative. Even the most negligible improvements in your life could be an act of creativity. When you are inventive, either in developing a new idea or using existing ideas in a novel way, you are creative.

A common misconception is that ideas just drop out of thin air in a "Eureka" moment, in the same way that the apple fell from the tree and gave Isaac Newton instant recognition of gravity! Even most metaphors about creativity are along the same lines: snap of a thumb, a light bulb moment, or a flash of insight. Most people do not realize how long Newton thought about the concept of gravity before arriving at his famous theory. The flash of insight occurred, but not entirely out of the blue as most people would seem to believe. We tend to highlight the moments of glory, forgetting all the steps leading to those moments! In reality, developing or using novel ideas or creating artistic work all happen in a process, with multiple iterations of combining and revising ideas. Creativity has a lot to do with combining existing ideas, disciplines, and cultures. Matt Ridley, zoologist and the author of *The Rational Optimist: How*

Prosperity Evolves, says, "Creativity is what happens when ideas have sex!" Effective brainstorming could organize and stimulate that process.

Being creative is rarely about coming up with an entirely new concept or idea. Instead, it has a lot to do with perceiving patterns, whether hidden or apparent, making connections between different facets of information, and eventually producing novel solutions to existing problems. Albert Einstein defines creativity as *"seeing what everyone else has seen and thinking what no one else has thought"*. But if you would often think up new and innovative ideas and don't bother to work on them, you are merely imaginative and not creative.

Creativity is not so much as trying one surprisingly new thing and finding out that it works. Instead, it is trying out many new things until you find one that works. It is the generation of multiple ideas until you stumble on something original and practical. To generate all those ideas, you have to allow yourself to step out of the box because how could you stumble on something original when you stay within the boundaries of un-originality? This is why brainstorming is a crucial element of creativity because brainstorming is essentially a deliberate effort to generate and evaluate as many ideas as possible. We will talk about the idea generation process in groups and on our own in Chapters 5 & 6 and discuss a wide range of brainstorming tools in Chapter 4.

Can We Improve Creativity?

Is creativity born, or is it made? Nature or nurture? Most people tend to believe that creativity is inborn; either you have it or do not. However, recent studies show that it is a little bit of both. In one of these studies, Dr. Beben Benyamin and his colleagues at Queensland Britain Institute reviewed various personal traits in a sample of roughly 14.5 million pairs of twins and published their findings in *Nature Genetics*. Their results showed that, on average, genetics and environmental factors almost equally control the twins' physical, mental, and emotional traits. Of course, the role of genetics and environment is not equal in all traits. But when it comes to being creative, a stimulating environment and following effective idea generation processes are as important as being intelligent, if not more important!

One of the misconceptions about creativity is that only brilliant minds could be creative. Perhaps that is true for Big-C creativity but not for the other three types. Creativity can be fanned, and individuals can be guided into their most creative selves. An excellent way to do that is through exposure to different ideas. After all, creativity is born out of an orgy of ideas having sex! This is where brainstorming comes into the picture. We will discuss many more tools to foster creativity in the following chapters.

Out of the Box Thinking

We are often encouraged to "think out of the box" when looking for novel ideas. But what is this "box"? The "box" is the border of our current knowledge. Inside the box, we feel comfortable and confident. The world inside our mental box is predictable and relatively well-known. Some people prefer not to think outside the box because it makes them feel insecure. The "box" is an imaginary safe space. Everything you know is inside that box, and there is no surprise or fear of the unknown because you know what's inside. It is comfortable, and who doesn't want to be comfortable? However, going outside the box could generate discomfort because it challenges everything you know and compels you to consider situations outside of your control or even your ability to predict. It is scary, and it can put your reputation at risk.

Now, thinking in the "box" isn't necessarily a bad thing. Rules and regulations form various boxes that help society remain functional. For example, the criminal justice system is a box, the Constitution is a box, and the traffic regulations form another box. It helps set the rules, keeps everyone in line, and establishes a system that benefits everyone.

So why do you have to think outside the box? Thinking outside the box is how people move forward. Before you can discover what you don't know, you must first establish what you do know. The box serves as a boundary between what you know and the things you haven't thought about yet. Everything we enjoy today is because there were people who chose to walk out of their boxes. Electricity, the internet, computers, GPS are all created by those searching for ideas and solutions not known before.

A famous example to practice out-of-the-box thinking is solving the "nine-dot problem." Maier first designed the nine-dot problem in 1930. It requires that nine dots arranged in a square be connected by four straight lines drawn without lifting the pen from the paper and without retracing any lines. Solving the nine-dot problem is possible by literally going out of the boundaries set by the nine dots.

Here is another story of how out-of-the-box thinking fosters creativity. A long time back, in a tiny village, a farmer ended up owing a large amount of money to the village moneylender. The old and ugly moneylender fancied the farmer's young and beautiful daughter and suggested the farmer a bargain: to let him marry the daughter in return for forgoing the debt. The farmer and his daughter both refused the proposal. So, the moneylender suggested letting Providence decide the matter. First, the moneylender would put a white and a black pebble into an empty bag. Then the girl will have to pick one out of the bag. If she picks a black pebble, she would have to marry him, and her father's debt will be forgiven. If she picks a white pebble, then she would not have to marry the moneylender, and the debt would still be waived. But if she refuses to pick a pebble, her father would be imprisoned. While the moneylender was putting the pebbles in, the daughter noticed that he deliberately put two black pebbles in the bag. So no matter what happens, the decision would be in his favor. Next, he asked her to pick a pebble. Imagine you are in her shoes. What would you do? You could try to expose the moneylender and all the complications that might follow. What the girl did was ingenious! She picked a pebble and, while looking at it, pretended to mistakenly fumble and drop it onto the bed of pebbles they were standing on. Then, she said, "But never mind, if you look into the beg for the remaining one, you will be able to tell which pebble I picked!"

The moneylender did not dare to accept his dishonesty, the girl saved herself of the misery and his father of the debt!

What is Brainstorming?

Brainstorming is not an entirely new concept. It was conceptualized by Alex F. Osborne as early as 1939, resulting from his frustration over his employees' lack of creativity. An advertising executive, he decided to experiment with different techniques to generate creative ideas via teamwork. Instead of allowing his employees to think in their offices independently, he called meetings like group thinking sessions where everyone could pitch in their thoughts. The term "brainstorm" was also coined by Osborne. He defined it as "using the brain to storm a problem." In 1953, he published his ideas and experiences on brainstorming sessions in his famous book *"Applied Imagination: Principles and Procedures of Creative thinking."* bringing the exercise and word into mainstream business.

While brainstorming was initially developed for idea generation in a group, it could be even more effective when used by individuals. We will discuss the group and individual brainstorming and their combinations in more detail in Chapters 6 and 7.

In group brainstorming, the participants try to find as many solutions for the intended problem as possible by building on each other's ideas. A facilitator often leads the brainstorming sessions while the participants present ideas in a free-flowing manner. The thoughts are often unfiltered and numerous. The facilitator's function is to collect and record all these ideas, promote the generation of more ideas, and essentially guide the team into focusing on the problem at hand. Nowadays, there are also various digital tools and apps to conduct brainstorming sessions in person and online. We will review some of these tools in Chapter 5.

Brainstorming is essentially an exercise of quantity rather than quality, with each idea recorded in one form or another. After collecting as many ideas as possible, the participants evaluate, combine and improve them. Often, it's not just about which one idea works but rather which combination of ideas may be used to address a specific problem. The same principles apply in individual brainstorming, i.e., first generating ideas while thinking freely, evaluating, and combining them afterward.

Why Brainstorming Matters?

The benefits of brainstorming have been phenomenal over the years, making it a staple of innovation and creative problem-solving in many companies. Here are a few examples of how brainstorming could benefit companies and individuals:

- Brainstorming sessions allow people to think without fear of being criticized. This paves the way for the generation of more "out of the box" ideas because no one will tell you that it is impossible, stupid, or too hard.

- The inputs from many people bring fresh perspectives into the mix, generating concepts that others may not have thought of. For example, men and women often see things from a different perspective. The same is true for the young and old.

- Brainstorming promotes the generation of a multitude of ideas within a relatively short time, increasing the chances of finding a brilliant idea.

- The process puts ideas out for study and contemplation by other participants. This allows the idea to be criticized, expanded, refined and accepted if it proves to be effective. After all, an idea is rarely, if ever, fully baked from the beginning. You have to look at brainstorming as a way people come up with ingredients for a brand-new recipe. The cooking does not start until all the recipes have been laid out on the table.

- It's perfect for team building since the final idea is often the result of collaboration and mixed concepts among the team members. This ensures that it is nobody's and everybody's idea! No one is trying to look and sound better than the others because, by the time the idea is fully baked, it's already a mixture of many ingredients. Who knows who put what on the table?

- Conclusions are reached by consensus. Everyone feels like their voice was heard during the meeting. This doesn't just help in solving problems, and it also creates a powerful team atmosphere.

- The session makes expressing and exchanging ideas easier, especially if the setting is informal and relaxed. In a well-managed brainstorming session, the participants will feel safe and not afraid of being judged. Everyone can express unorthodox ideas; who knows what comes out of those seemingly unusual ideas!

- Individuals can follow the same rules to have a brainstorming session independently or before joining group brainstorming. In this case, having the right mindset is essential. Individual brainstorming could happen with a more flexible structure. For example, there are fewer or even no time restrictions. Some people might think about the idea for a new business, a song, or a movie for years before getting started with it.

Brainstorming Examples

Brainstorming can be used in various ways, depending on the context, available tools, and the results you want. For example, big companies like Google and Amazon employ various brainstorming tools to develop new business ideas or improve the existing ones. If such big companies rely on brainstorming to help them move forward, why shouldn't you give it a shot? Let's review some brainstorming examples in both professional contexts and everyday life.

Amazon Working Backwards

Amazon is undoubtedly one of the biggest brands nowadays, offering a wide range of products and services. One of these products is *Amazon Echo* smart speakers and screens. According to Amazon Echo's Vice President, Mike George, his team used a "working backward" brainstorming technique to develop the product idea. This method of brainstorming starts with a pretend press release. The team pretends to draft a press release for a product, even though it is undeveloped or hasn't been conceptualized yet. The press release essentially touts all the benefits of their brand new "product", ignoring all possible technical impossibilities that might come with its development. The press release is the embodiment of all their aspirations for the product. From there, the group works backward to develop the technology needed to turn those aspirations into reality. According to Mike George, they also write FAQs to identify possible questions they would be getting about the product. By doing this, they can create a much clearer picture of what they want to build.

Google Ventures Brainstorming Sessions

Google Ventures is an independent arm of Alphabet Inc. It provides the initial and growth capital financing to bold new companies in life science, healthcare, artificial intelligence, robotics, transportation, cybersecurity, etc. Google Ventures needs to generate tons of ideas and be ahead of the curve for tech development. So how do they do it?

In 2016, Jake Knapp shared his decade-long experiences working for Google Ventures in his bestseller book *Sprint*. Knapp mentioned the "Note-and-Vote" technique as the primary brainstorming tool used by Google Ventures to quickly and efficiently generate ideas. It works by allowing the team members to write down as many ideas as possible with ZERO self-editing. Once all ideas are collected, the whole team reviews them for a fixed period (for example, five minutes), picks their favorite idea, and shares it with others after the review time is over. The selected ideas will be written down on a whiteboard so that everyone can see them. After some evaluations, people vote for the best ideas. A "decider," which is often chosen in advance, can make the final decision once the ideas have been whittled down to just a few concepts.

This might seem like a long process, but it only takes 15 to 30 minutes to accomplish it in practice. Even better, the note-taking process means that participants won't have to worry about other people seeing and criticizing their ideas. They can go as crazy or as elaborate as they wish before eventually picking out their favorite portions. This helps create a safe space for all members to stretch their imagination. Of course, the process can take more or less time or happen differently in each session. For example, the "decider" might not be needed or could be skipped if the group prefers a more democratic approach. But the main elements always remain the same.

According to Jake Knapp, such techniques help teams work individually and as a group simultaneously, therefore creating better results instead of just using one or the other independently. With this method, teams can decide on company names, product names, and even where they want to hold the next Year-End Party!

Mall Mapping

Believe it or not, brainstorming is often done when trying to map a brand-new mall. Do you think all those kiosks, stores, and restaurants are placed there at random? Of course not! The placement is often finalized after ideas are thrown around, ensuring that the complete arrangement encourages people to buy. Some spots are known for getting better traffic and, therefore, better chances of sales. This is why you'd often find large shops hogging the outlets right in front of the entrance, where tons of people will pass by, and they can be instantly seen!

Buffet and Retail Outlets

The brainstorming setup extends even to buffets and retail outlets. Ever wondered why the staples are put at the back of the shop? It's so you'll have to pass by all the other products first before getting to the items you need. Who knows, by the time you get to the eggs, you've already placed other items into your shopping cart. Somehow, stores have brainstormed their way into a layout that encourages people to buy to earn more with each shopper. It also took some brainstorming before Walmart started putting their high-price items in front of the store. You're not going to buy their expensive products, but the exorbitant cost of their flat-screen television makes it seem as though whatever you buy inside will be small in comparison.

And what about buffets? Well, it's the same principle but on the reverse. Brainstorming created a buffet layout that encourages clients to get cheaper filling food offerings like pizza, spaghetti, and rice. It's arranged in such a way that by the time you pass by more expensive items, lobster, beef, chicken, you're already pretty full and will eat less of those.

Here are other small snippets of creativity probably resulting after people went through brainstorming process in search of a solution for a problem:

- How can we help stray animals with food and water? One developer created a vending machine that chucks out food and water for pets. To get the goods, people can put in plastic bottles as payment. This hits two birds with one stone by encouraging people to recycle and feeding the stray animals!

- Some stores offer two types of shopping baskets; for example, black ones mean the person using them prefers to shop alone, while a red basket means they'd like to be assisted during shopping. This makes things easier for the attendants and keeps the introverts and extroverts equally happy during their shopping!

- Have you ever sat in the dentist's chair and got bored looking at the ceiling? So one dentist decided to put a "Where's Wally" design on the ceiling so that patients can stay entertained while being treated. Sure, a television up there might be better, but a "Where's Wally" could be more fun!

- You're coming home after shopping, and your hands are packed to the brim with grocery items. How do you control the elevator? Most people use their elbows or put some of the bags down if the elbows don't work. Well, how about an elevator fashioned with extra controls on the bottom – for your feet!

- Have you ever decided you don't want to buy an item after all and just shove it in the nearest aisle? It's not surprising to find snacks in the noodle section or even frozen hotdogs in the liquor section! How about a store with "landing racks" where every "change of mind" item can be left to be put back by the staff? This is quite common in libraries where books are put back in "landing tables" so the librarian can return them to the correct section.

Of course, those are just some real-world examples of how brainstorming can be used to solve problems and make our lives easier. We can also use brainstorming in everyday life situations to generate ideas and make decisions. Here are some examples:

- Someone's birthday is coming up, and you want to decide on the best theme for the party.

- The garden needs a revamp. How do you design everything in a fresh look?

- New furniture's coming in, and you're not sure how to place everything. It's time to brainstorm a new home design.

- Need to do some creative makeup for Halloween? Brainstorm something spooky and attention-grabbing!

- You can brainstorm the best way to teach your child how to clean, eat healthily, or improve their study habits.

- You can even brainstorm how to motivate and discipline yourself into sticking to a diet, an exercise routine, or even a skincare routine.

- Perhaps you want to film a TikTok video and want to find creative ways to showcase your ideas?

Literally, any problem in this world can be approached through brainstorming, especially if you are looking for a novel and creative solution or something you haven't tried before. How to brainstorm effortlessly and effectively is what this book is all about!

What Can You Discover in This Book?

The discussion you've read is just the tip of the iceberg. After all, we're still in the Introduction, and there is still a lot you can learn about brainstorming, how it can unleash creativity, and how it can pave the path towards bigger and better business developments.

The book is structured in the following three main sections:

- *Part 1: Understanding creative brainstorming (Chapters 1,2):* This part of the book talks about creativity and how to foster it, the definition, history, and applications of brainstorming, and how to use brainstorming for creative problem-solving.

- *Part 2: The process of creative brainstorming (Chapters 3,4):* This part discusses the "process" or the "approach" to find new and innovative ideas. We still do not talk about the "tools" (i.e., the methods and techniques). Here the focus is not on the outcome of the ideation but on how that outcome is reached.

- *Part 3: Brainstorming for personal and professional problem-solving (Chapters 5,6,7):* Here, we will talk about the "tools" and "techniques" to organize ideation sessions (both individually and in groups) while following the processes and approaches explained in the previous chapters. After introducing a wide range of brainstorming techniques, the application of these techniques for problem-solving in personal and professional life will be discussed. The advantages and pitfalls of group brainstorming, the benefits of individual brainstorming, organizing effective brainstorming sessions, and combining individual and group brainstorming in various phases of the ideation process are among the main talking points of this part.

Brainstorming is like a comprehensive tool to help you improve your personal and professional life. If you're stuck in one form or another...,

Maybe with the right name for your baby to be born soon?...

Or with a marketing campaign idea?...

Or maybe with design ideas for your upcoming renovation project?...

Hopefully, this book will give you all the necessary ingredients to become a better brainstormer! I wish you a lot of fun and learning by reading this book. Thank you for your time, and good luck!

The Art of Brainstorming

Chapter 1: Fostering Creativity

George Bernard Shaw says, *"Imagination is the beginning of creation. You imagine what you desire, you will what you imagine, and at last, you create what you will."* Creativity is the act of turning imagination into something tangible, just like when Bernard Shaw turned his thoughts into masterful plays. As mentioned in the introduction, this book is written to answer one question: "How to generate novel ideas for creative problem-solving?" Let's see how we can answer this question if we follow Shaw's advice. If finding a creative solution is your desire, first, you need to imagine that solution. Imagination is passive; it's envisioning an idea. You should also will what you've imagined. In other words, you should be willing to work on your imagination to refine and improve it actively. Eventually, you can bring your idea to life and create a novel solution.

This chapter takes a closer look at the process of being creative, very much like what George Bernard Shaw laid out. Throughout this book, we will discuss creativity in the context of problem-solving rather than artistic creativity. With a good understanding of creativity, we will be ready to embark on our journey to master the art of brainstorming. Let's get started!

The Essential Ingredients of Creativity

What exactly do you need to be "creative"? In other words, what are the main ingredients for the creativity recipe? Tina Seelig, the author of *"InGenius: A Crash Course on Creativity"*, nicely explains these ingredients. Here is a brief overview:

1. *Knowledge:* Everything starts with what you know. Some would even say that your knowledge establishes the boundaries of your box. Interestingly, the knowledge you have doesn't necessarily have to be confined or related to the problem you're trying to solve. Knowledge from one field could be a great source of inspiration for solving problems in an entirely different area. For example, scientists could not figure out how Neanderthals used a piece of polished rib bone as a tool in their everyday life. However, when shown to a leather craftsman, it was uncovered that the bone tool was used for burnishing leather. Thus, to spark creativity, knowledge in various fields can sometimes be as helpful as in-depth knowledge in a particular area.

2. *Imagination:* To prompt new ideas, one requires purposeful imagination. What does that mean? Purposeful imagination is the ability to picture something new or innovative, usually drawn from an existing pool of knowledge, to address the intended problem. Purposeful imagination has a lot to do with connecting seemingly unrelated concepts and combining and reshaping existing ideas.

3. *Attitude:* Of course, one must have the right attitude to get the ball rolling. Attitude encompasses motivation, passion, persistence, and the mindset that you can solve a particular problem. It is the "will" Bernard Shaw was referring to, the governing force that turns the idea into action and then persists until it is reshaped and polished into a novel solution.

4. *Habitat:* While some people can generate and execute ideas anywhere, most require a habitat that encourages a creative mindset. For example, a company and its work culture create an environment or a habitat that can promote or punish creativity. The family or school culture has the same function. The habitat must be such that it is conducive to creativity, allowing for failure or removing any penalty that you would typically associate with a loss. A good example would be Google's *20 Percent Project*, which allowed Google employees to spend 20 percent of their time working on creative projects of their choice while still being paid

Once all of the ingredients mentioned above are present, creativity has a favorite ecosystem and will flourish.

Furthermore, we can accelerate the process by applying specific thinking techniques that act as fertilizers. These techniques are especially important when creativity is pursued by teams or groups of people, such as in most professional contexts. The creativity of a team highly depends on how the individual members interact with each other. Do they synergize or not? Even the most creative people benefit from interactions with their peers and friends. For example, Pablo Picasso created his best works of Cubism after years of collaboration with his lesser-known colleague Georges Braque.

The Essential Thinking Techniques for Creativity

Creativity is not just bringing new ideas to this world. It is also about optimizing those ideas and reducing their complexities. The same applies to brainstorming. Although brainstorming starts with generating as many ideas as possible, it doesn't stop there. We will see throughout this book that brainstorming happens in cycles of idea generation and idea analysis. In this section, we will review three essential techniques to enhance creative thinking. These techniques are related and often used together to get the best possible outcome. Furthermore, they form the basis for brainstorming techniques discussed in the following chapters.

Out-of-the-Box Thinking

We briefly discussed out-of-the-box thinking in the Introduction. So let's talk more about how to get our minds out of the thinking box! Before we continue, it is essential to note that our goal is to find innovative solutions which offer added value or make life easier. Out-of-the-box thinking is not just about generating unorthodox ideas. We can all do that! The goal is to find novel ideas with added value. Sometimes, there is a good reason for staying in the thinking box. For example, imagine having square wheels under cars! It's an unorthodox idea but is also dumb! Eventually, every idea, whether generated by thinking in the box or outside of it, should make sense and offer some value.

As mentioned in the Introduction, the box is the proverbial boundary between everything you know and those you don't know. But what creates the box? In many cases, the box is a combination of the learned and the inherited. Genetics, for example, plays a massive role in how the box is perceived. How would a deaf person define the boundaries of their box? What about someone who is blind?

Our environment, education, social circle, successes, and failures also help set the boundaries for the box. Sometimes, we even create these boxes for ourselves, such as when we develop habits. Think about it; what's the first thing you do when you wake up? What's the route you always take to work and back? These are all part of our boxes, and some may argue that staying within those boxes is the logical thing to do. Since the box is known, it is safe and predictable. You can design your life according to what is safe and secure. On the flip side, going outside the box means you'll have to delve into uncharted waters. This means leaving the known for the unknown. There's no shame in admitting that this is a scary thing to do. However, thinking out of the box is a necessary step towards creative problem-solving. In the words of Thomas Jefferson: *"If you want something you've never had, you have to do something you've never done."*

You might wonder why the box even exists if it's detrimental to creativity. The fact is that the box is there not just to keep you safe but to make everything else easier. This is perfectly logical, and the brain is hardwired to create the proverbial box and keep you within it.

In *"The Power of Habit,"* Charles Duhigg explains that habits are formed by combining a trigger, an action, and a reward. It can be something as simple as having an alarm as your trigger, preparing coffee as the action, and having a nice hot cup of coffee as a reward. What does this have to do with creativity? Psychology literature suggests that our brain is a pattern-finding machine. It operates subconsciously and settles into familiar grooves and patterns. Once the habit is formed, an activity becomes practically effortless, like keeping your balance once you've learned how to cycle. If you create the proper habits, everything in your life could be a regular part of a routine that you don't even think about or worry about anymore. This includes working out, preparing dinner, walking the dog, going for a hike, and various other activities. Once a task has turned into a habit, it requires very little work for the brain to process it.

Now, this pattern-seeking behavior of the brain can also extend to how we think. Exposure to the same thought processes leads us to follow them over and over again. So, on the one hand, habits make our lives easier, but on the other hand, they can be a hindrance to creativity. The key here is to distinguish what activities can be kept in the box and which should be taken out, studied, and then put back. Stepping out of the box sounds easier in theory, but in practice, it's not. Once you put it into practice, it becomes evident that thinking outside the box is very challenging. After all, how are you supposed to think about things you don't know?

Perhaps the best way to enhance out-of-the-box thinking is to enlarge our thinking box! It is usually easier to search for new ideas based on our existing knowledge than to create new ideas from what we don't know. A great way to expand our thinking box is to constantly expose ourselves to new ideas, meet people from different backgrounds, visit new places, expand our social and professional network, have new experiences, learn new skills, and so on. Most people think out-of-the-box thinking is trying hard and forcing ourselves to generate brilliant ideas that we never knew about before. This is simply a misconception. Out-of-the-box thinking is more about challenging assumptions and going beyond our predetermined mental boundaries. This will give you a different angle on how to attack specific problems. Think of your knowledge and experiences as blocks of lego. The more blocks you have, the more shapes you can build. Creativity is how you combine those lego blocks. Having a more extensive collection of blocks (i.e., having a larger thinking box) provides you with more building blocks to play with and enhances the chances of coming up with new combinations. The ability to make new and innovative combinations using existing ideas is the most critical creative thinking technique. We will discuss this in the next section.

Productive Thinking

Growing up, we are often forced to perceive the world through categories. We categorize apples as fruits, broccoli as vegetables, cake as dessert, and so on. While this way of thinking helps us function, it can also be detrimental to creativity. Productive thinking is a form of out-of-the-box thinking by combining different concepts to create a brand new one. For example, take "staycation," a combination of the words STAY and VACATION. It sounds like two opposing concepts, right? After all, the whole point of going on a vacation is leaving your home. Today, "staycation" is a concept used to describe people going on a vacation within their locality. For example, you've decided to tour a neighboring city for the weekend, or perhaps you're given one whole weekend alone in the house, and you can do whatever you want. Essentially, the concept creates a situation where you're not going far, but you enter the mindset of someone who is actually on vacation. How did anyone come up with this idea?

The book *"Imagine: How Creativity Works"* written by Jonah Lehrer, explains that the best way of blending concepts takes the input of several people. First, you have to bring colleagues together to share information and address a common question. Perhaps present the problem and allow them to "hibernate" on it before allowing them to jump into each other's pool of thoughts.

The core concept of productive thinking is "connect and combine," i.e., to find two different concepts and mash them together to create something brand new. The key is to find a commonality between the two concepts. It doesn't have to be a huge one, just a tiny point where the two ideas seem to intersect.

Let's go back to the term STAYCATION. If you look at the root concepts of this brand-new word – they're complete opposites of each other. They come from two completely different categories. However, what happens if you expand your vision of the two terms? If you tilt it a little or change the angle of how you look at these two concepts – is there a way for them to fall into a similar category? Yes! They're both ways of spending your free time. Staying at home and going for a vacation are both ways in which people pamper themselves or re-energize themselves during the off-work days. Once you find that small dot where the two concepts intersect, it's not a giant leap to figure out how you can somehow combine the two ideas into a unified one.

Another great example of productive thinking in practice is the Japanese art of "Chindogu," meaning "unusual tool" or the art of useless innovations. Chindogu items are gadgets for everyday life that anyone can understand their use. It is primarily a humorous problem-solving practice to solve minor everyday problems. What makes Chindogu clever and exciting is that it is on a very fine line between "What a crazy idea!" and "This might actually work!".Some examples are the dust mop that babies can wear to clean the floor while crawling, the chopstick fan that cools your while on its way to your mouth, a broom and dustpan slipper attachment, or a toothbrush finger attachment. You might argue that these Chindogu creations are all funny if not useless, but the important thing here is the intent behind the inventions. No matter how weird most ideas are, some will prove valuable or become the basis for developing a functional product.

Divergent-Convergent Thinking Cycle

The terms "divergent" and "convergent" thinking were first coined in the 1950s by J.P. Guilford, an American psychologist who focused on human intelligence. He defined divergent thinking as a problem-solving approach that considers multiple possible solutions. He observed that creative people tend to show this type of thinking quite often. Guilford proposed the concept of "divergent thinking" with the following characteristics:

- **Fluency:** The ability to produce many ideas or problem solutions in a relatively short period,

- **Flexibility**: The ability to develop a variety of approaches to solve the same problem,

- **Originality:** The ability to produce new, original ideas,

- **Elaboration:** The ability to organize the details of an idea and improve it with multiple iterations

In a creative thinking cycle, divergent thinking is when a free-flowing or a spontaneous flood of ideas occurs. We will see later in this book that divergent thinking is an integral part of the brainstorming process, but not all of it.

This is the other side of the thinking coin, according to J.P. Guilford. Convergent thinking happens when different ideas are brought together to find the one best solution for the problem as if the problem at hand is a multiple-choice question with only one correct answer. While divergent thinking encourages multiple possible solutions, convergent thinking is culminated by deciding on one single answer. Thus, convergent thinking is more suitable for well-defined problems with a clear scope.

Convergent thinking is a handy tool to analyze and rank ideas, but not so much to predict, imagine, compose, and create them. The world of convergent thinkers is black and white, with no shades of grey. But we all know that our world does not always work that way! Creative problem-solving often involves working with incomplete data, uncertain outcomes, and unpredictable influences in the real world. Hence, it is impractical and often impossible to aim for one best solution. For example, imagine you want to predict the outcome of multiple marketing campaigns, decide on how to invest in different financial climates, or plan your career steps. It would be challenging (if not impossible) to find the best course of action for such questions. That's why divergent and convergent thinking are often combined in cycles.

Convergent thinking is all about managing to choose the correct answer in a multitude of possibilities. This is precisely how our education works, and for the most part, students' grades are measured based on their ability to practice convergent thinking properly. Divergent thinking, on the other hand, is all about coming up with different ideas. A good game that measures divergent thinking includes asking how many uses they can find for a specific thing. For example, there are two people. Each one is given a large piece of cloth measuring six by six feet and asked this question: How many possible uses can you find for this cloth?

One person may say that the cloth can be used as a blanket, as a bedsheet, or perhaps as a towel and then stop there. However, another person could go on and on about its possible uses. It could be a blanket, a sheet, a fort, a cape, a table cloth, a curtain, a cheesecloth, a sling, a muumuu, and a wide variety of other possible uses for the product. The divergent thinkers usually pave the way for new ideas that other people may not instantly think about.

It is the use of both divergent and convergent thinking that leads to creative and relevant ideas. These two thinking strategies are akin to brainstorming sessions in the generation and conceptualization of an idea. However, they can't happen at the same time because they need different mindsets and approaches. In Chapter 4, we will discuss how to utilize the divergent-convergent thinking cycle for effective brainstorming.

Domain Specificity of Creativity

A major misconception about creativity is that it is universal and encompasses all walks of life. The assumption is that if you take a creative writer and have them direct a movie, their creativity will extend to film-making. In other words, a creative person will be creative regardless of the circumstance you put them in. However, more and more studies are proving the opposite; that creativity is domain-specific rather than domain-general. Therefore, a creative writer would not necessarily be a creative filmmaker, and vice versa.

Dr. John Baer, a professor at Rider University and a recognized author and researcher in creativity, has done some of the best studies in this field. His focus has been on whether creativity is domain-general or not, i.e., whether creativity in one domain or field extends to other disciplines. His extensive studies showed that there is no single cognitive function associated with creativity. That means there is no "general" attribute among people, which makes it possible to spread creativity from one area to other areas

Let's put it this way. A creative writer cannot, by simple association, be a creative director, a creative singer, a creative painter, or a creative cook. For creativity in a particular domain to prosper, one has to be knowledgeable about that subject in the first place. For example, a chef can be creative with their dishes because they understand the different aspects of food, flavoring, spices, and cooking methods. Dr. Baer suggests that creativity is like "expertise" and is highly domain-specific. If you ask a professional chef to create a brand-new dessert, their "expertise" makes it easy for them to be creative. How about we ask our creative chef to use his creative streak to design a prom dress? How do you think that would turn out? Domain-specific creativity dictates that the chef won't be so good when designing a dress because his expertise is in a completely different field. He doesn't know enough about cuts, fabrics, and colors to be confident in experimenting with clothing design.

What about polymaths? Polymaths are experts in multiple domains. Examples are Leonardo da Vinci, Benjamin Franklin, Nikola Tesla, Hellen Keller, Aristotle, and Avicenna. Doesn't their existence prove that there's such a thing as domain-general creativity? Do polymaths prove a common creativity streak can exist in people so that a single person can be creative in multiple domains? Well, polymaths are the exceptions that prove the rule.

They are so rare that their existence demonstrates how difficult it is to be creative in multiple domains. It requires a very high degree of intelligence and motivation to attain expertise in various subjects – and that's what polymaths are.

Interestingly enough, even with polymaths, the scope of the expertise tends to overlap. Let's look at Leonardo da Vinci, famous for his drawings, paintings, sculpture, inventions, and writings. He was even skilled in anatomy, geology, astronomy, botany, cartography, paleontology, music, and mathematics. Clearly, he is a polymath, but there are obvious overlaps if you look at his domains of expertise. His expertise in the drawing made him an efficient painter and sculptor. Since he was an excellent mathematician and artist, he can quickly design and build new mechanisms, turning him into a great engineer.

Domain specificity does not say that a person can only be creative in one domain, only that creativity in one domain is not predictive of creativity in another domain. Anyone can be creative, but not in any domain. The underlying reason for the domain-specificity of creativity is that creativity often happens in divergent-convergent thinking cycles. In other words, many ideas are created in the first phase of the cycle (for example, in a brainstorming session). In the second phase, those ideas are analyzed and judged for their value and impact. Without expertise, the convergent part of the thinking cycle cannot happen. So although we might have a bunch of ideas, we would not be able to assess and rank them.

Earlier in this chapter, we discussed three thinking techniques to foster creativity. These were divergent-convergent thinking, out-of-the-box thinking, and productive thinking.

Let's take a step back and see how to apply these techniques knowing that creativity is domain-specific. The techniques mentioned above encourage us to have multiple thinking stages (each with their specific characteristic and mindset), look at problems from different and even unusual perspectives, and combine and connect existing ideas. When people get together in a brainstorming session, they could combine their expertise from various fields, challenge assumptions, build on each others' ideas, and generate novel and better ones. That is why brainstorming is such an effective tool for creative idea generation.

In one of his scientific papers published in 2015 in Roeper Review, Dr. Baer beautifully explains collective thinking, similar to what happens in the brainstorming process, is so pivotal to creative problem-solving:

"The domain specificity of creativity means that interdisciplinary thinking, interdisciplinary collaboration, and interdisciplinary creativity are even more important than one would assume if creativity was domain-general. One cannot simply transfer expertise from one domain to another, unrelated domain. One's expertise in Edo art will be of little use in reading computed tomography scans, nor can one simply transfer skills, motivations, personality traits, or other creativity-relevant factors from one domain to another. Domain specificity does not mean that projects that demand interdisciplinary solutions disappear, but it does help us understand the need for multiple skills, insights, and expertise in solving many problems."

In the following chapters, we will dive deeper into applying various thinking techniques to create novel ideas, the tools to make brainstorming useful and pleasant, how to avoid the common pitfalls of brainstorming, and many more topics. Now that you have a more clear picture of creativity, it's time to talk about brainstorming!

CHAPTER 2: UNDERSTANDING BRAINSTORMING

By now, you should have a better picture of what creativity is and how to foster it. In the remaining chapters of this book, we will discuss specific tools and processes for idea generation and refinement practically and straightforwardly. Brainstorming will be our primary tool and a vehicle to implement what we have discussed so far. This chapter will examine brainstorming, its origins, workings, advantages, and drawbacks in detail. You will see why brainstorming is such a powerful tool. And a perfect approach to implement the divergent-convergent thinking cycle and the productive thinking

History of Brainstorming

We can trace the origins of brainstorming back to the late 1930s when Alex Faickney Osborn started to use the technique to generate creative ideas by his employees. Osborne was the co-founder and advertising executive of the US advertising agency BBDO. Frustrated by his employees' inability to generate creative ideas for ad campaigns and have fruitful exchange of ideas, he started experimenting with new basic principles of conducting effective business meetings. He summarized his ideas to the following four principles:

- Focus on producing as many ideas as possible

- Avoid criticizing ideas

- Try to share wild and unconventional ideas

- Build upon each others' ideas

These principles were designed to stimulate free, creative, and collaborative thinking without fear of being judged. By 1939, he could already see a noticeable improvement in the quality and quantity of the ideas produced by his colleagues. Osborn saw that following these principles generated more ideas, giving rise to more valuable ideas. He continued to refine the principles and eventually published them in his 1942-book *How to Think Up*. Osborn described the process as *"a conference technique by which a group attempts to find a solution for a specific problem by amassing all the ideas spontaneously by its members"*. He initially used the term "think up" to refer to the creative idea generation process. But Osborn indirectly created the word "brainstorming" by referring to the thinking process as *"using the brain to storm a creative problem."*

In 1953, Osborn published a new book titled *Applied Imagination*. This is where the term "brainstorming" really got some traction. In *Applied Imagination*, Osborn formulated the rules and principles of approaching creative problem solving, and many are still relevant today! Since its birth in the early 1940s, brainstorming has continuously spread, especially in professional environments. Every day, many brainstorming sessions are held worldwide to generate novel ideas in many areas, from consumer research and marketing to advertising campaigns, insurance policies, new product development, etc.

Brainstorming has proved to be a valuable tool. The caveat is that to do it correctly. This book aims to help you know brainstorming in detail and learn how to conduct pleasant and productive brainstorming sessions with your colleagues at work, the family and friends at home, or on your own! So, let's dive deeper into the brainstorming principles as laid out by Alex Osborne decades ago and refined and revised ever since.

The Principles and Practice of Brainstorming

At the heart of Osborn's creative thinking process lay two principles:

- *Deferring judgment*: so that people are concerned more about generating ideas rather than defending them, especially unorthodox ideas,

- *Reaching for quantity*: hoping that high quantity increases the odds of having high quality.

Osborn formulated these principles to stimulate free, creative, and collaborative thinking. In addition, he encouraged his team members to engage in some role-playing; first, withholding all criticisms and letting the ideas flow, then analyzing, refining, and combining them. Most people focus only on the first brainstorming role; thinking freely and generating as many ideas as possible. That is where the drama and the excitement are! They tend to forget the second role; analyzing, screening, and refining ideas. This oblivion afflicts brainstorming in much the same way as if the role-players feel embarrassed and abandon the play altogether. We will talk about the pitfalls of brainstorming later in this chapter.

Osborn also established four rules of brainstorming based on the principles mentioned above. The first three rules were to focus on quantity, withhold criticism, and encourage out-of-the-box thinking. These were the rules for the first part of the brainstorming role-play. The intention was to leave no stones unturned to find ideas, even if they sound unrealistic or low-quality, hoping that one or more diamonds in the rough are found. The fourth rule was to analyze the ideas and find the one with high potential. This was the rule of the second part of the game.

In addition to the four rules, Osborn emphasized the importance of clearly defining the question in need of creative answers. This is like setting the scene for the play before starting. Albert Einstein is quoted as having said, *"If I had an hour to solve a problem and my life depended on the solution, I would spend the first 55 minutes determining the proper question to ask... for once I know the proper question, I could solve the problem in less than five minutes."* Good preparation is vital for any group discussions, especially a creative ideation session. Which problem are the participants trying to solve? Which question are they going to answer? It would make a huge difference if the question is "How can we build a bridge on the river?" or "How can we cross the river?" The first question limits the scope of the brainstorming to only one of the possible solutions. The second question still focuses on the end goal but sets a different scene.

To kickstart a successful brainstorming session, it is essential to define the problem in a way that is not vague and does not strangle creative thinking.

Let's take a closer look at each of the four original brainstorming rules proposed by Alex Osborn:

1. *Quantity over Quality*: Osborn viewed brainstorming as a tool to generate as many ideas as possible, believing that the number of ideas must be massive to create a large pool where effective ideas can be fished. The excessive amount of ideas also opens the possibility of merging, developing, and refining each item to meet a common goal. Osborn was encouraging divergent thinking by this rule.

2. *Criticisms must be withheld.* The fear of criticism or rejection is often enough to stop ideas from being voiced out. Osborn knew this from the experience, and he consciously advised withholding the outpour of criticism until the session was over. This is an extension of Rule #1 since the mind is set free and generates a broader range of ideas without the fear of coming up with a "dumb" idea. The sheer mass of ideas also enhances team ownership, as it can often be tough to pinpoint the exact generator of the concept once the brainstorming session is over.

3. *Encourage out-of-the-box thinking*: Again, an extension of the previous two rules, wild and unorthodox ideas are not only allowed but also encouraged. There are no crazy ideas in the divergent thinking phase of brainstorming. All ideas, once uttered, deserve merit no matter how obvious it is that they will not work. The previous two rules, i.e., pursuing

quantity and withholding criticism, provide the foundation and the safe space for out-of-the-box thinking. The point is although most wild ideas prove to be impractical, even if one of them turns out to be directly or indirectly valuable, this stage of the brainstorming session could be a game-changer.

4. *Combine, refine, and improve ideas:* Osborn knew that generating many ideas is not enough to solve the problem at hand. That's why he formulated Rule #4 to incorporate a convergent thinking phase as the final stage of the brainstorming process. Ideas are like legos piled on top of each other, scattered all over the place. Once all the ideas are laid out, it's time to use them as building blocks to make fascinating objects. Convergent thinking is at the heart of this stage, and criticisms and insights are welcome. By twisting, fusing, and refining the ideas generated in the previous steps, an even bolder one might emerge. So, Rule #4 contributes to Rule#1 by creating even more ideas, and to Rule #3 by enhancing the odds of finding out-of-the-box solutions.

As we will see later in this chapter and throughout the book, the definition and practicalities of brainstorming have evolved over the decades since Osborn's time. But the principles and the main rules are still valid.

In addition to the rules and principles, Osborn introduced concrete guidelines on how to conduct brainstorming sessions. Here are some examples:

- Choose participants with knowledge about the subject matter so that they can assess the quality and the relevance of various ideas.

- Choose participants with varying amounts of experience in the intended subject to look at the problem from various perspectives.

- Inform the participants about the specific problem before the session to focus on idea generation during the session.

- Do not mix participants from various levels in the company's hierarchy to let ideas flow freely.

- Train the participants on the rules and principles of brainstorming before the session.

- Have a facilitator before and during the session to increase the odds of fruitful and inspiring brainstorming.

- A company executive or leader should analyze the ideas generated in the brainstorming session to choose the viable ones and decide on the next steps.

- The ideal size of the brainstorming group is 5 to 12 people.

Osborn even paid attention to the color and furniture of the meeting room, its facilities, etc., to create a warm, relaxed, and inspiring atmosphere and let creativity flourish. In BBDO brainstorming sessions, a stenographer recorded all discussions. After the session, a company executive or leader sorted and analyzed the ideas. In 1956 alone, BBDO had 47 continuing brainstorming panels, held 401 brainstorming sessions with a total output of 34,000 new ideas, and generated 2000 ideas considered worthwhile. The brainstorming process worked marvelously and transformed BBDO from a struggling agency to a flourishing one within a decade.

The brainstorming process was so successful in BBDO that it soon became a trending word in both the business world and academia. By 1958, some of America's most prominent corporations and universities utilized brainstorming to generate creative ideas. These included the Aluminum Corporation of America, Du Pont, Better Homes and Gardens, Kraft Foods, General Electric, Harvard Business School, Massachusetts Institute of Technology, and the University of Southern California.

The Evolution of Brainstorming

The widespread adoption of the brainstorming process by American corporations and universities made it an interesting research topic for academia. Donald Taylor, Paul Berry, and Clifford Block of Yale University conducted one of the earliest and most prominent studies on brainstorming in 1958. They wanted to prove that group brainstorming is not as effective in idea generation as individuals thinking up on their own. They chose a group of ninety-six Yale students and asked half of them to independently develop creative solutions for the intended problem. They divided the other half into groups of four headed by a lead and asked them to brainstorm creative ideas. The research team found that the individuals generated more ideas than the brainstorming groups and concluded that individual brainstorming is more productive than group brainstorming.

Before we give up on Alex Osborn's method, let's pause and take a closer look at the Yale study. The experimental design of that study had some significant deviations from Osborn's process. Here are some examples:

- The group size was too small.

- The Yale students were not representative of the experience level of the participants in a real-life brainstorming session.

- The students did not have any ownership of the problem (i.e., no skin in the game).

- They received absurd problems for brainstorming, such as the issues faced by people who suddenly grew an extra thumb!

- There was limited facilitation before and after the brainstorming sessions.

Despite all of the design flaws mentioned above, the Yale study significantly impacted the evolution of brainstorming. The biggest one was to introduce the notion that individual brainstorming is better than group brainstorming. This was, of course, a significant departure from Osborn's process. Later studies comparing the productivity of the group and individual brainstorming focused more on generating high-quality ideas than a large number of ideas, which was another deviation from Osborn's methodology. Gradually, the definition of brainstorming was expanded to include both group and individual creative idea generation. Individual brainstorming was defined as a process in which one person thought up ideas on their own.

Brainstorming has remained a hot research topic ever since its introduction. After the Yale study, many more researchers investigated various brainstorming aspects, mainly to find the most effective and productive brainstorming practices. One of the best resources to summarize the vast body of brainstorming research was published in 1998 by Scott G. Isaksen of the Creative Problem Solving group in Buffalo, New York. He reviewed ninety studies on brainstorming conducted since the Yale study in 1958 until 1998. Isaksen found out that many researchers departed from the Osborn group brainstorming process and could not replicate the settings used by Osborn, for example, in the following aspects:

- Using nominal groups instead of real working groups,

- Quality of facilitation before and during the group brainstorming,

- Level of training the participants in the brainstorming groups received.

Because of such differences, the researchers discovered it was hard to prove or disprove whether Osborn's group brainstorming process was superior to individual brainstorming or not.

Interestingly, if we look closely at Osborn's brainstorming process model, we can see that it already included individual brainstorming. Osborn recommended isolated ideation both before and after a group brainstorming session. As mentioned before, he advised the participants to know and understand the problem before attending the session. After the brainstorming session, a company executive or a leader sorted and analyzed all the ideas and selected the viable ones. Therefore, as described in the Yale study, we can say that isolated ideation occurred before and after the BBDO group brainstorming sessions.

It seems like the body of research done on brainstorming since the mid-20th century could not reach any conclusive answer on which brainstorming practice is better in generating creative ideas. But, interestingly, they have eventually converged on Osborn's method! This is not surprising because Alex Osborn built his ideation method on solid principles and clear, practical rules. Moreover, he focused on what worked the best for his business and his colleagues. He paid attention to the entire ideation process, involving the right people, informing them about the problem before the session, letting their minds roam free during the session, and analyzing the ideas afterward.

Alex Osborn's brainstorming process has passed the test of time, and we can still learn from it. The key is to keep the principles in mind and stimulate creative thinking for your specific situation.

Creative thinking is the prerequisite for effective brainstorming. Regardless of what brainstorming approach or method you utilize, it would help if you think creatively. Chapter 2 briefly reviewed three essential techniques to enhance creative thinking. We can use these techniques, some introduced after Alex Osborn's time, to streamline and strengthen the brainstorming process. Before we go down to the nitty-gritty of brainstorming sessions, we will first discuss creative thinking techniques in more detail in the following two chapters. Imagine you have the perfect conditions for brainstorming in a group or on your own. How can you improve the chances that you will think out-of-the-box and generate lots of ideas? Is it possible to define principles for the creative ideation process? The following chapter will answer these questions!

CHAPTER 3: PRODUCTIVE THINKING

"The unexpected connection is more powerful than one that is obvious."

Heraclitus, 6th-century B.C.E.

Albert Einstein and Charles Darwin are among the most influential scientists of the last two centuries. Their ideas have changed the way we understand the world around us. So what kind of thinking ends up creating such groundbreaking ideas? You might think Einstein and Darwin were geniuses, and ordinary people could not reach their level of ingenuity. While this might be true in terms of intelligence, it is not exactly the case for their creativity. In 1907, Einstein realized that he could not explain the behavior of an accelerating object using his theory of special relativity. To solve this problem, he looked at it from different perspectives. He imagined someone sitting in a room in outer space accelerating at 9.8 meters per second (the same as earth's gravitational acceleration). Einstein then asked himself how light would behave in an accelerating room. He found out that light would bend in the accelerating room since gravity and acceleration are equivalent. But he could not find the correct mathematical expressions for his finding. In 1912, Einstein's mathematician friend, Marcel Grossman, introduced Einstein to tensor analysis and helped him explain his theory mathematically correctly. It took Einstein three more years of hard work before he published his theory of relativity in 1915. So even a genius like Einstein had to approach the problem correctly and get help from other disciplines.

Charles Darwin and his famous book *Origins of Species*, published in 1859, have a similar story. The theory of evolution did not occur to Darwin without any precedent in scientific history. The raw material for his theory had been known for decades. Geologists and paleontologists already knew a great deal: that life had existed on earth for a long time, many species had become extinct, species were related to each other, and so on. Darwin's genius was to formulate all those evidence in a theory that explained the evolution of species from a common ancestry through a plausible mechanism, which he called "natural selection."

Darwin was not even the only person to discover natural selection. The British biologist Alfred Russel conceived the same mechanism independently. Darwin and Wallace both found out that if an animal has certain traits that help it become more resistant against the elements or breed more successfully, that animal will have more offspring. As a result, those traits will become more common over time. Interestingly, Darwin and Wallace found inspiration for their theory from various sources: from economics to pigeon breeders. For example, when they studied *Essay on The Principle of Population,* a book published in 1797 by the British political economist Thomas Malthus, they hypothesized that plants and animals should face the same population pressure as humans experience. They believed only those species which are more adaptable could survive such pressures and thrive over time. Darwin also spent a great deal of time with pigeon breeders while trying to formulate natural selection. He observed how breeders selected certain pigeons to reproduce and develop a neck ruffle. Darwin argued that nature does the same by selecting individuals better suited to their environmental conditions and, given enough time, even creates new body parts for them.

What characterizes the thinking strategies of Einstein or Darwin? The examples mentioned above already demonstrate some of these strategies, such as looking at the problem from different perspectives, gaining inspiration from various sources, building on the existing knowledge, etc. How about the thinking strategies of highly creative artists such as Picasso? What can we learn from them to brainstorm better?

This chapter focuses on the ideation process, i.e., the approach to get new and creative ideas. We will analyze how geniuses and creative artists think and how to clone their thinking strategies. Once you know more about the ideation process, you can take that process and use it to facilitate your brainstorming.

Productive and Reproductive Thinking

One of the most important creative thinking strategies is productive thinking. The first person to introduce productive thinking was the German psychologist Otto Selz in the early 1930s. Max Wertheimer, another German psychologist, adopted and expanded Otto Selz's ideas and published his book *Productive Thinking* in 1945. Wertheimer compared and contrasted productive and reproductive thinking. He associated reproductive thinking with repetition, forming habits, and staying within familiar intellectual territory (i.e., our mental box).

On the other hand, he explained that productive thinking revolves around insight, restructuring of the problem, and combining different concepts to produce new ones. Creativity theorists such as Tim Hurson and Michael Mickalko have revisited productive thinking from various perspectives in recent years. They have published their views in *Think Better: An Innovator's Guide to Productive Thinking, Cracking Creativity,* and *Thinkertoys.*

Reproductive thinking relies heavily on experience. Some people call it "inherited" thinking because it applies old solutions. When you use a stick to get something stuck on a tree or get something floating in the pool, you are thinking reproductively. In other words, you are reproducing solutions that you have already used before. It's a quick approach, and most people rely on this method to solve a problem quickly or take low risks. We go with the tried and tested thinking process to have a high degree of certainty about the outcome. On the other hand, productive thinking goes beyond past experiences and tries something new to solve a problem.

Is it wrong to think reproductively? Not at all. In a setting where patterns are pretty common, reproductive thinking can speed up the whole problem-solving process. Moreover, reproductive thinking is very efficient when a new problem could be solved with an old solution. However, it can also impede creativity because of its fixedness.

Productive thinkers reframe and rethink the problem and consider it from different points of view. They combine existing ideas and merge them into new ones. This problem-solving approach is prevalent in geniuses like Einstein, Darwin, or Galileo, as mentioned at the beginning of this chapter. Although it is challenging to define a standard procedure or method to replicate the productive thinking of geniuses, we can still learn a great deal about the attributes of the process that they follow, as we'll discuss later in this chapter.

In *Cracking Creativity*, Michael Michalko argues that productive thinking is very similar to the biological evolution of species. They both create multiple alternatives from the exiting options, but only a few of them survive the selection process. In biological systems, natural selection decides which species will survive. In productive thinking, the analysis, intellect, and facts select the best idea. The same process of creation and elimination happened in Alex Osborn's brainstorming process; they first created as many ideas as possible in the brainstorming session and then sorted, analyzed, and down-selected them. The only difference is that the selection process in productive thinking or brainstorming is not as random as natural selection.

Biological evolution occurs when individuals of a species have sex, combine their genes, and create new individuals inheriting characteristics of their parents. Creativity works in the same way. It happens when attributes and aspects of specific ideas are combined and refined to form new ideas. In *The Rational Optimist*, the zoologist and science writer Matt Ridley explains how most progress in human history happened due to the meeting and mating ideas. In other words, as Ridley calls it, "ideas having sex with each other."

Ideas Having Sex

The meeting and mating of ideas is the cornerstone of productive thinking. The brainstorming process pioneered by Alex Osborn is, in essence, an organized procedure to let ideas meet and mate. The goal of all brainstorming sessions is to have many ideas, let them have sex with each other, and produce new and hopefully creative ideas. So, how exactly can we give our ideas a better sex life? Let's review four common ways to encourage the mating of ideas.

1. Combining Random Objects

If you're looking for a brand-new idea, combining random objects could be a great way to go. You can create two lists of random things, pick one object from each, and think about different combinations and associations those two objects might create. For example, what happens if we put a Christmas tree and fish together? They might seem completely irrelevant. But an avid fisherman came up with the idea of using unsold Christmas trees to create a new habitat for fish in nearby lakes. Volunteers (primarily local fishermen) took the trees and secured them to the lake bed. The newly denuded branches were soon covered by algae, attracting aquatic insects and a lot of fish. Soon, the fishermen had their private fishing locations!

2. Combining Unrelated Ideas

It's also possible to allow dissimilar objects to have sex with each other. One excellent real-life example is combining "toilet" and "playing games." In the early 1990s, a Dutch designer named Aad Kieboom figured out that he could solve the age-old urine splashback problem by putting tiny "target" pictures on toilet bowls. As a result, men are more likely to aim for that target, thereby limiting the chances of urine splashing everywhere. First introduced at Schiphol Airport in Amsterdam in 1999, etching the image of a fly in the urinal decreased the splashes by more than 80 percent.

Another idea brought about by combining unrelated objects is the basketball hoop trash can. By putting basketball hoops on top of trash cans, people are encouraged to throw their trash properly. They tend to ignore the trash can a lot less and get entertained while keeping the area clean.

There are countless other examples of creative inventions out of unrelated concepts and objects. Here are some of them:

- Combining a bell and a clock to get an alarm clock.

- Combining a trolley and a bag to get a suitcase with wheels.

- Combining copier and a telephone and get a fax machine.

- Combining an igloo with a hotel to get an ice palace.

3. Combining Problems

Coming up with a single answer for two or more problems is another example of productive thinking. Think of FedEx's forklift scale, which allowed to weigh items during transportation. Digital spoons use the same idea. They weigh the ingredients while scooping up. The spoon has an LED screen to show the weight creating a convenient way of following recipes.

4. Combining Talents

The fourth way to approach the sex of ideas would be by combining talent in people. This is where polymaths gain the advantage as their knowledge in different areas allows them to make more combinations. In a group setting, it makes sense to bring together people with diverse backgrounds. The combined ideas of the group add together, thereby allowing them to come up with an answer that no one can develop independently. If the group is too big, you can divide them into smaller groups, making sure each group combines different talents.

Combining different and even totally irrelevant ideas is the cornerstone of creativity. Imagine the following scenario to demonstrate how to implement the ways mentioned above in an everyday life context. You are a teacher and would like to explain and demonstrate conceptual blending to your students. The basic principle for you is to encourage sex between dissimilar concepts and use creative thinking to work out the dissonances.

Here are some examples to spark the discussion between your students:

- Making the U.S. Constitution more portable,

- Modeling a human face with regular polygons,

- An engineer designing a pop song,

- Huckleberry Finn in the Hunger Games,

- Making novels more interactive,

- Mathematical formulas in the Mona Lisa,

Each example contains a dissonance that might sound weird and ridiculous at first glance. Maybe some students will show adverse reactions. That's, in fact, a good sign because it shows that your students recognize the cognitive dissonance. Most probably, the first feedbacks will be something like, "Why do we need the US Constitution to be portable? Or How about just carrying a copy of the Constitution? Or Who wants to carry a copy of the Constitution after all?

These reactions show that the connection between the students and the US Constitution is not that strong because they do not see the point of having the Constitution close to them. When you carry something with you (such as photos of your loved ones in your wallet), that object is most probably significant to you.

The next step is to push your students to think of different ways of approaching the question? In other words, ideas other than carrying a copy of the Constitution around. Here are some examples:

- Making Bill of Rights T-shirts or watch,

- Translating the US Constitution to modern-day English,

- Composing a Constitution rap song to allow people to listen to the Constitution while being entertained,

The cognitive dissonance in the original questions encourages the students to engage in the brainstorming process. It feels fun and playful to look for new ideas and could enhance fun and lively atmosphere.

Thinking Like a Genius

You don't have to be a literal genius to develop something new and innovative. You just have to think like one. In *Thinkertoys*, Michael Michalko looks through history to find the common threads in how geniuses develop their creative ideas. If you believe the common attribute among science, art, or literature icons is their extraordinary intellect, you're mistaken! As we discussed in Chapter 1, creativity is not the same as intelligence. Michalko explains how geniuses work more following the laws of biological evolution. In other words, they borrow ideas from various disciplines and combine them to create many different alternatives. What sets geniuses apart from ordinary people is, first and foremost, their ability to consider the problem from multiple angles and to create more combinations of ideas to solve that problem.

Nature works in the same way. It creates a vast diversity of species and then allows natural selection to decide which species will survive. The outcome of natural selection is not known in advance. But the goal is clear: survival. Indeed, some plants and animals adapt in exciting and unusual ways to survive. For example, while Alaskan Wood Frogs freeze their bodies to survive the harsh winters, some Antarctic fish make anti-freeze proteins to survive the frigid waters of the Southern Ocean. So, nature managed to find two totally different solutions for a single problem!

It would be naive to think that geniuses do not rely on their intelligence, hard work, focus, perseverance, and luck. But not everyone with these attributes would become a genius. The secret ingredient to add to the above list is how geniuses re-think the problems, combine ideas, and make connections and attributions. The way Einstein conceived his theory of General Relativity beautifully portrays this secret ingredient. When Einstein was coming home from a visit to his friend Michele Besso in May 1905, he remembered riding a streetcar in Bern, the Swiss capital, where he spent two years looking back at the city's famous clock tower. He then imagined what would happen if the streetcar traveled away from the clock tower at the speed of light. Since light could not catch up with the streetcar, the clock would appear stopped.

On the other hand, his own clock would beat just fine. He realized that the clock would appear stopped since light could not catch up to the streetcar, but his own clock in the streetcar would beat normally. He suddenly realized that time is not a fixed concept and can beat at different rates throughout the universe, depending on your speed. Thousands of people should have traveled the same streetcar and seen the same clock tower without giving the whole experience a second thought. Only

Einstein observed the world around him in a different way and kept thinking until he finally tapped into "God's thoughts."

Geniuses like Einstein can take something mundane like the view from a streetcar, combine it with their existing knowledge and create something that the world has not thought of before. Of course, it might be difficult for most of us to reach the level of Einstein's genius. Nevertheless, we can learn a great deal from them and maybe clone some of their thinking strategies. The surprising fact about these strategies is that they're not exactly extraordinary. They are, in fact, simple and elegant. Let's take a closer look at six thinking strategies common to geniuses throughout history.

1. Considering Problems from Different Angles

Looking at a problem from different perspectives has always been a hallmark of creativity. For example, Leonardo da Vinci believed that the first way he sees a problem is often biased, so he would take the time to look at it from different perspectives. His understanding grew with each new way of looking at something, allowing him to create an intelligent combination of various perspectives.

How can we practice this in real life?. How about challenging or even reversing our assumptions? For example, let's say you're looking for ideas for a new kind of hotel. Here are some basic assumptions about hotels:

• Hotels have a building.

• Hotels have rooms.

- Hotel rooms have beds.

What happens if we reverse these assumptions?

- Hotels have no building.

- Hotels have no rooms.

- Hotel rooms have no beds.

How about generating ideas to bring the reverse assumption into reality?

- Hotels without building: an outdoor hotel, a hotel inside a cave, a treehouse hotel, etc.

- Hotels without rooms: hotels with "cocoon" shaped accommodations, a hotel with indoor tents, etc.

- Hotel rooms without beds: rooms for temporary (to take a shower, rest on a couch, work, etc.), a hammocks hotel, hotel with a big shared lounge, etc.

You can continue this process for a set amount of time or until you reach a satisfactory amount of new ideas.

2. Making Novel Combinations

This is an extension of the first strategy. Approaching a problem from different perspectives and considering the opposites of what we assume usually create novel combinations. For example, let's take a good look at Sesame Street. In this show, two categories of characters appear in the show: the puppets and the humans. Before Sesame Street became such a popular show, people thought that kids' characters should be either purely puppets or purely human. They categorized the performers because they were afraid that children would be confused about which is real and which isn't. When creators decided to abandon the societal mindset, they came up with Sesame Street and had a huge success.

3. Thought Visualization

Galileo, one of the most impactful scientists of the renaissance and a recognized polymath, would often use diagrams to trace, understand, and explain his ideas. This is no different from the mind mapping we do today; creating a visual approach to problem-solving. In contrast, many of Galileo's contemporaries stuck with using mathematical and verbal methods, somehow limiting their perception of the subject. Instead, try drawing diagrams, flowcharts, graphs, sketches, etc., and observe the flow of information as you make your ideas pop out of the page. Albert Einstein always said he thought in pictures: *"Words do not play any role in my thought; instead, I think in signs and images which I can copy and combine."*

4. Churning Out Lots of Ideas

Geniuses tend to churn out more ideas to have a higher chance of one sticking and becoming successful. It's like throwing darts onto a target. If you throw enough darts, one is bound to hit the bull's eye. So, while most of us may not have an IQ comparable to that of Leonardo da Vinci, we can at least try to come up with as many ideas as possible. Mozart wrote 600 pieces of music in his lifetime, and not all of those are exceptional. Edison and his associates worked on at least three thousand ideas for an efficient incandescent lamp.

5. Forcing Relationships

Sometimes, ideas don't have to fit perfectly together to work. Ideas don't come out as perfectly formed pieces of a puzzle. Often, it's important to chip away at the corners so that two different items match up and solve the problem.

A good example would be Samuel Morse, one of the developers of the Morse code to send telegraphic messages over long distances. He was thinking about the best way to transmit messages across the US, primarily worrying about generating a signal strong enough to send a message from one side of the country to another. Finally, the answer came to him while watching how horses were exchanged in a relay station. He figured out that transmission is possible if the signal is boosted along the way. Nowadays, we use cell towers to "boost" our messages by allowing them to bounce from one site to another.

6. Preserving and Processing Your Ideas

Suppose you try everything we've discussed so far in this book, and you get lots of ideas to solve a particular problem. The chances are high that most of those ideas are half-baked and not as brilliant as you would like. So you need to develop certain habits to use those thoughts as precursors for breakthrough ideas. Here are four examples of such habits:

- Make sure to write your ideas down as early as possible, including the half-baked ones.

- Do not expect to hit the idea jackpot immediately. Instead, give your ideas time and let them mature.

- Give your ideas time to develop and mature, share them and get feedback as often as possible.

- Play with your ideas: combine and subtract them, turn them upside down, refine and revise them, etc.

Einstein once said, *"It's not that I'm so smart, it's just that I stay with problems longer."* Studying geniuses' thinking strategies and habits show how crucial this "staying with the problem" is. Dwelling on the problem requires everything we've discussed so far: approaching the problem from various perspectives, writing and organizing ideas, allowing ideas to have sex, getting feedback, analyzing and refining, etc. Of course, every problem and its context are different. But having the right mindset, discipline, and thinking tools will take you far!

Thinking Like a Designer

The outcome of brainstorming is not known in advance, the same way that artists, poets, writers, designers, or architects do not know exactly how their final product will look like. Instead, they start with a concept or draft, do iterations, refine the draft, and repeat this cycle until they're satisfied with the outcome. In that sense, brainstorming is like designing ideas. Although you might have some facts and data about the problem to be solved and even a well-defined question, you do not know the outcome of the brainstorming session. So, you need the flexibility and the mindset of a painter or an architect

Earlier in this chapter, we compared ideation with the evolution and natural selection processes in biological systems, how nature creates various living organisms and lets them evolve and adapt over many generations. This section will turn to the world of art for new inspirations and thinking strategies. The main question is what we can learn from highly creative artists such as Picasso to become better brainstormers.

1. Volume Matters More Than Perfection.

Remember what we mentioned earlier about Mozart? That Mozart wrote 600 pieces of music in his lifetime, and not all of those are exceptional. Most people know Picasso for his Cubism paintings. But he did a lot more than Cubism. Picasso was very prolific. He created more than 1800 paintings, 1200 sculptures, 2800 ceramics, 12000 drawings, and countless other artwork such as prints and tapestries. Perhaps not all of these works were great. But staying in action and creating so many art pieces were the stepping stone for Picasso to produce his masterpieces. The same principle applies to brainstorming: first, create a large pile of ideas, combine and refine them, and finally, keep going until you have a brilliant one.

2. Adopt Good Ideas and Add Your Twist.

Many people think that creative geniuses like Einstein or Darwin were born with great ideas. But as we discussed earlier in this chapter, they used the existing knowledge, got help from other fellow scientists, received feedback, and succeeded after years of hard work. Picasso is widely quoted to have said, *"Good artists borrow, great artists steal!"* Perhaps what Picasso meant was to draw inspiration from as many resources as possible and not just imitate.

There is no such thing as "original" without knowing what's been done before. You cannot become an original thinker, innovator, or artist without learning what the old masters have done. Do you think Picasso started right off the bat with his Cubist masterpieces? No! He started off learning how to reproduce realistic imitations of great artists like Rafael.

Originality does not mean you never use existing ideas. On the contrary, originality is constantly inspired by what is known. The key is to add your spin, interpretation, and intuition and create something new to this world.

3. Think Like a Child.

Do you remember when you were a child and could imagine, draw, and build without limitation? We grow up, become adults, and that creativity dies. Our teachers, parents, and society tell us to follow specific rules. Then, we become adults and end up struggling to think outside those rules.

Engaging in brainstorming to express yourself and have fun like a child, with the same curiosity and freedom of mind, makes a big difference. But as an adult, how can we keep our creative inner child alive, for example, at work? If you want to break the rules, color outside the lines, and think out of the box, you first need to know the rules. This is where once more, the domain-specificity of creative thinking becomes essential. Think of Picasso again. He first learned how to create hyper-realistic paintings. Then he started thinking if a human head consists of elements like mouth, ears, eyes, and a nose, why should they always be in specified locations? What if they're moved around?

That's how he started with his Cubism paintings. Picasso is quoted to have said, *"It took me four years to paint like Rafael, but a lifetime to paint like a child."*

4. Think and Talk Visually

Verbal presentation of ideas is good. Visual representation, like the sketches of a painting or drawing, is even better. Graphics and illustrations help make the concept more tangible, giving others a better chance of seeing what the originator means. Nothing explains what a painter, architect, or sculptor has in mind better than a sketch or model. The same is true for innovative ideas and prototypes. Displaying ideas helps build the web between the information. Thoughts are like trees; they tend to branch out and make connections. Someone's rough idea could lead to another person's good idea and perhaps a brilliant one. Having information on display creates an excellent jump-off point to inspire the others, thereby creating a solid connection between all participants.

This chapter reviewed the creation process of nature, geniuses, and artists. Despite the differences between the biological processes of evolution and natural selection and the creative process of artists and geniuses, there is a great deal of commonality between them. The common thread is the divergent and convergent cycles of creation and elimination. In other words, first, a large number of possible alternatives are created before refining and narrowing down to the most viable ones. Interestingly, Alex Osborn's brainstorming process follows the same concepts and the same divergent-convergent thinking cycle.

Ideation is usually not linear, as shown in a diagram. Instead, you have to analyze and critique, go back and forth in divergent-convergent cycles, combine and refine existing ideas, add your twist, and adapt and expand them. Nailing the right question can lead to the development of the right answer

Now that we've examined creativity and creative thinking strategies and know more about the history and foundations of brainstorming, it is time to talk about the nitty-gritty of brainstorming. In the upcoming chapters, we will review a wide range of group and individual brainstorming techniques, the pitfalls to avoid, and the best practices to follow.

A Short message from the Author

Hey, hope you're enjoying the book. I'd love to hear your thoughts! Many readers do not know how hard reviews are to come by, and how much they help an author.

I would be incredibly thankful if you could take just 60 seconds to write a brief review on Amazon, even if it's just a few sentences!

Please scan the QR code to share your thoughts:

Your review will genuinely make a difference for me and help gain exposure for my work.

CHAPTER 4: BRAINSTORMING TOOLS AND TECHNIQUES

Since it was conceptualized in the early 1940s, brainstorming has always been an essential technique for creative idea generation. Over time, different variants of traditional brainstorming are developed, primarily to remove the drawbacks of group brainstorming. Nowadays, there are many tools and techniques to facilitate and organize brainstorming sessions. Whether you are going to brainstorm face to face or online, brainstorming tools could help spark more ideas and gather and manage them efficiently.

What are brainstorming tools and techniques? We are not talking about basic items such as post-it notes, pen and paper, or a whiteboard. Instead, we will discuss strategies to approach the problem from different angles, frame the right questions, enhance collaboration between the participants, improve engagement, and generate as many great ideas as possible. One of the important topics in this chapter is the online or virtual brainstorming tools. Such tools are precious with the recent trend of remote working. They allow for digitizing the entire idea generation process and working as a team without losing your creativity.

Chapter 2 introduced traditional brainstorming, its origins, principles, and rules in detail. The current chapter will review the most common shortcomings of conventional brainstorming, present ten major variants of traditional brainstorming, and X other idea generation techniques based on conventional brainstorming principles. Later chapters will utilize these tools to discuss setting up efficient group or individual brainstorming sessions.

Shortcomings of Traditional Brainstorming

Does this sound familiar to you? Your manager or team lead asks you to participate in a chaotic exercise billed as a brainstorming session. They encourage you to "think out of the box", and then await you and the rest of the team to have a eureka moment and generate brilliant ideas that never occurred to anyone before! Unfortunately, too many conventional brainstorming sessions are plagued with a lack of planning and organization, ill-defined goals, divergent thinking only, and a few dominant speakers with others silently sitting through most or all of the meeting. In many organizations, having a brainstorming session is more like "a tick in the box" affair with no tangible outcomes. Most brainstorming issues relate to group interactions and behaviors and are prevalent in team business settings in corporations and start-ups. Company politics could also create a barrier against the free-flowing generation of ideas. In other words, some participants might focus on what the boss likes to hear.

As we discussed in Chapter 2, brainstorming sessions require proper planning and facilitation. Otherwise, they might yield little or no genuine results. Let's review some of the most common pitfalls of conventional brainstorming. We will later discuss how to avoid them and which new techniques are available to address these pitfalls.

1. Production Blocking

Production blocking happens when one individual blocks or inhibits other participants during a group discussion. For example, if the brainstorming group consists of six people and one presents their idea, then the other five are "blocked" and unable to provide their own creative input. As a result, they may not have time to think up new ideas or might get distracted and forget their ideas before they have an opportunity to share them. Most people cannot think up new concepts while listening to someone else presenting their idea. Production blocking becomes more of an obstacle as the size of the brainstorming group grows. We can solve this problem by allowing individuals to write their ideas before joining the group discussion, thus preventing any "blocking" during the session.

2. Evaluation Apprehension

This is the typical fear that comes with voicing out the wild, crazy, and unconventional ideas. Even before people blurt it out, they think: what will others think? This fear effectively stops people from going off tangent, diminishing the number of ideas that the group could evaluate at the end of the brainstorming session. Thus, even with the brainstorming rules, the fear of being criticized can be a powerful deterrent for creative problem-solving. Evaluation apprehension results in self-censorship and reduced efficiency of the brainstorming process. This problem, however, is often solved by new techniques such as Brainwriting.

3. Topic Fixation

Someone managed to come up with a great idea, and somehow, everyone seems to conclude that no other idea can be better. Hence, even during the brainstorming session, everyone's mind gets fixed on the particular topic, essentially narrowing the possibilities for other ideas. It is as if the participants think, "We've already come to a consensus, so why pitch another idea?" This problem has been spotted and solved through different brainstorming techniques like the Step Ladder or Brainwriting.

4. Dominant Personalities

Have you ever met someone who has such a powerful personality? They can walk into a room, and everything they say seems to be rock solid. Once they leave, however, the idea doesn't seem extraordinary, but you find yourself agreeing to it because the person who came up with the concept is compelling. Strong personalities take control at the beginning of the session and maintain dominance afterward.

Unfortunately, traditional brainstorming sessions often suffer from this pitfall. Some members are just so convincing that they can easily sway the group's opinion and block all avenues for creative thinking. Proper facilitation and techniques such as brainwriting or mind-mapping could address this pitfall.

5. The Lack of Anonymity

Another prevalent fear among members of brainstorming groups would be the face-to-face setup of the session. You're not just afraid that people will ridicule your suggestion. You're worried that they'll also criticize you for coming up with it in the first place. Despite the team approach to brainstorming, there are instances when ideas, good or not so good, can be easily traced to a particular person. Therefore, it's common for many people to withhold their unorthodox ideas because they could be so easily attributed to them. This can be solved using anonymous brainstorming techniques such as brain-netting.

6. Too Many Cooks

Do you know what they say about too many cooks? Well, traditional brainstorming can also suffer from the same problem. A group containing too many people can easily create a chaotic atmosphere that makes it tough to organize ideas. Everyone can be talking at once, and even if they're not, it will take a long time for each person to be able to pitch in. Limiting the number of people participating or choosing who should participate depending on their expertise can help solve this problem.

Over the past decades, many companies and individuals have tried to address the shortcomings mentioned above. The outcome is a wide range of brainstorming variants with different approaches and focus areas. The rest of this chapter will discuss twelve new brainstorming techniques. These techniques share the core brainstorming principles of group synergy and divergent-convergent thinking cycles, each with its unique twist.

Questionstorming

Developed by Hal Gregersen from MIT Sloan School of Management, questionstorming is like brainstorming for questions. The underlying idea is that fresh, open, and honest inquiry creates novel insights and makes it easier to venture into uncharted territories. Questionstorming could be more effective than traditional brainstorming because it allows people to analyze the problem by asking questions rather than looking for answers. As mentioned earlier in this chapter, evaluation apprehension is a significant pitfall of conventional brainstorming. People feel pressured when they are trying to find the "right" answer. When they are worried about getting that brilliant "right" solution, they tend to hold back for fear of looking stupid.

Questionstorming offers many advantages in comparison to conventional brainstorming. The first one is that since you do not know the right answer, you won't feel obliged only to ask the questions that seem right. This sense of freedom creates a more engaging atmosphere with more opportunities to find innovative ideas. The other advantage of questionforming is that it could redefine the initial question. Do you remember the example from Chapter 2 on how it would make a massive difference if the question is "How can we build a bridge on the river?" or "How can we cross the river?" If the brainstorming session starts with building a bridge without anyone inquiring about the exact problem at hand (i.e., crossing the river), the outcome might not be ideal. The sole focus on asking questions stops the participants from rushing to an answer or getting fixated on a possible solution. This will eventually expand the scope of ideations and Finally, asking questions is fun, and everyone can do it.

Gregersen built his questionstorming methodology on two critical rules:

1. Only questions are allowed: The facilitator should redirect any participant who tries to suggest solutions or respond to others' questions.

2. No preamble or justifications are allowed: This is to avoid framing a question and guiding people to see the problem in a certain way.

He also proposed certain principles to guide questionstorming sessions. Here are some examples:

- Use traditional divergent thinking techniques (such as making random associations) to unlock new questions.

- Start with simple questions and then shift to more cognitively complex ones.

- Questions should not put people on the spot, develop an atmosphere of fear, or be aggressive.

- Multiple short rounds of questionstorming are usually more effective than a single long session.

- Questionstorming should start with an emphasis on the quantity of the questions. Analysis, sorting, and prioritization come later.

Over the years, Gregersen has developed a standard four-step questionstorming methodology for both individual and group ideation sessions. He has used the methodology with hundreds of teams from many companies, including Chanel, Danone, Disney, Fidelity, Salesforce, and countless non-profit organizations and individuals. Here is how questionstorming works:

Step 1: Setting the Stage

Everything starts with a statement as opposed to a question. For example, when using brainstorming, you open the discussion with this question: "How can we increase our customers?" With questionstorming, however, the opening statement is often like, "We need more customers." Do you see the difference?

Step 2: Listing Questions

Once you've decided on a statement, the next step is to start asking relevant questions. This is done in the same way ideas are listed during the brainstorming process. You give participants a set amount of time to list as many questions as possible in response to the particular statement. For example, let's say the original statement is: we need more customers. The questions following this statement could be:

- Who are our customers?

- What's the demographic of our customers?

- Can we expand our customer base?

- How many customers do we have now?

- How much more can we add to our customer base?

- What makes our customers like us?

- Can we support more customers?

The beauty of this method is that asking questions is often easier than finding answers. There's no pressure to think hard and dive deep into your knowledge base; you just have to be curious. A good way of conducting this step is by giving participants around 10 minutes to develop as many questions as possible. However, you can try interrupting them at the eight-minute mark and ask them to come up with ten more questions. This technique helps make sure that they've exhausted all possible questions instead of just taking it easy for the last few minutes or so.

Step 3: Tweaking the Questions

When tweaking the question, try to change it from an open one to a closed one and vice versa. A closed question is any question that's answerable with a yes or a no. An open question requires a bit more explanation or elaboration. So, in the above example, the closed question is: Can we support more customers? If we open up that question, it would be: How can we support more customers? Flipping the questions from an open to a closed one also helps set the stage for follow-up questions. Let's retake a look at the above example. Can we support more customers? If the answer is YES, then "How can we support more customers?" is the logical next question. But what if the answer is NO? If the answer is negative, then proceeding to the open question isn't necessary. Instead, a new open question may pop up, such as: "How can we improve our ability to support more customers?"

By slightly changing the wording, you get a completely different view of the question. This technique also helps to have more questions and, therefore, widens the scope of the ideation.

Step 4: Sorting and Prioritizing the Questions

Coming up with the questions may be seen as the divergent thinking phase of the process. This is when you line up all your thoughts. With prioritizing the questions, you're now in your convergent thinking stage. At this point, you'll need to pinpoint the questions that usefully reframes the problem and could potentially open a new pathway. It would be best if you analyzed the selected questions to see why they are impactful or important. Finally, commit to pursuing at least one new pathway based on your selected questions. This is the part where you can start answering those questions.

The parallels between the brainstorming technique of Alex Osborn and the questionstorming methodology of Hal Gregersen are fascinating. They approach the problem from two very different angles but following almost identical principles. Questionstorming could be done individually or in a team, face to face, or online. The principles and the stages will remain the same. Only the process needs a little bit of adaptation to fit the context best.

What-If Questions

"If you don't know what you would do if you could do whatever you wanted, then how on earth can you know what you would do under constraints?" That was what Professor Russel Ackoff from Wharton School, University of Pennsylvania said. "What-If" questions make us think. They get us out of our current situation and take us to a world where anything is possible. It's an excellent way to warm up our minds for creative thinking or generate ideas that we never thought of before.

"What-If" questions provide the opportunity to get into the mindset of someone unfettered or free from the box. If you want to stretch out your "out of the box" thinking muscles, asking what-if questions would be an excellent way to get started. Such questions are to get the team "ready" for thinking outside the box. Here are some "what if" questions:

- What if we had an unlimited budget?

- What if children make all our decisions?

- What if we let our users design the project?

- What if we can sell this product in a different market?

- What if we sell all our products directly through the internet?

These fantastical questions are a good jump-off point for insights that hopefully will lead to creative ideas. The beauty of this technique is that it helps you envision seemingly unrealistic circumstances enough to propel you into thinking about the actionable steps to approach those circumstances. It opens the doors to the imagination, and with the imagination flung open, you could break the box open.

"What if" questions are like warm-up moves, making them an excellent exercise for a team that is new to brainstorming. Also, if they're having a hard time coming up with innovative ideas, asking "what-if" questions can bring them into a more creative mindset. Like all other brainstorming techniques, it would be better to spend some time on the front-end planning and facilitation of the What-If sessions or conversations.

Mind Mapping

Invented by Tony Buzan in the 1960s, mind mapping effectively creates a visual representation of your ideas. This does not mean creating drawings or caricatures of concepts. Instead, the focus is on finding connections between the thoughts organized around a core topic or question. Mind mapping looks like a flowchart, albeit the links might not be as organized as in a conventional flowchart. Creativity and association play a significant role in how the concepts and ideas are connected. The visual representation in mind mapping could also include different shapes, colors, and designs.

Mind mapping is a visual exercise that helps you see the connection between different aspects of a problem. For example, if a company is trying to save on operational expenses, a mind map could help trace connections between various items that expend money. In a printing shop, for example, paper consumption could be a substantial cost driver. What factors contribute to paper use? How much paper is being wasted every day? How can you decrease the waste of paper? Are there days when paper consumption seems to be more than usual? Why? What happens on those days, and how can you control them? These are all things that mind mapping can help organize and narrow down by making connections and representing them visually.

Typically, this method works better if:

1. You're trying to find connections and relationships between different concepts and ideas.

2. You want to combine different ideas.

3. You want to create a structured approach when planning a topic or assignment.

Here are the typical steps for mind mapping:

Step 1: Choosing a Medium

Start by trying to figure out what you'll be using for mind mapping. For example, you can use pen and paper, a whiteboard, a blackboard, post-it notes, or a tablet or laptop with mind mapping apps such as Simplemind or MindNode.

Step 2: Placing the Core Topic in the Center

To start, put your core topic in the very center of the blank page. Everything you write afterward should be somehow related to this core topic. For example, let's say your ultimate goal is to lose weight healthily. Put "healthy weight loss" in the center of your blank page.

Step 3: Finding Subtopics

The next thing to do would be figuring out the subtopics under weight loss. What are the essential components of losing weight? First, of course, there's diet and exercise. But what else? For example, there's the amount of sleep you get, your stress levels, and the position in the menstrual cycle for women. You can also put in your current weight and your goal weight. Next, place the subtopics around the core topic and connect them with lines. The next step is to add more ideas related to the subtopics.

If you're focused on "exercise" as a subtopic, you can think of various forms of exercise to lose weight. For example, walking ten thousand steps a day, cycling to work and back, dog walking, etc. How much physical activity do you have now, and how can you do more?

List down any related ideas under the subtopic. For example:

- Taking ten thousand steps or an average of five miles per day as a good starting point.

- How do I measure my steps per day? Install a step counter app in my phone or use a fitness band.

- What exercises can I do?

- Walking, jogging, cycling, swimming, HIIT, yoga, etc.

Next, you can list related ideas when it comes to diet:

- Ideal calorie consumption per day is at two thousand calories for men and one thousand five hundred calories for women.

- How do I measure my calories?

- Different diet regimens like Ketogenic Diet or Intermittent Fasting.

Step 5: Connecting and Finding Relationships

Finally, it's time to find relationships and connections between the different subtopics. The goal is to see how one point is related to another and how you can use that information to achieve your end goal. For example, you can make a connection between diet and exercise. Food is all about taking calories, and activity is all about burning them. The key to weight loss is taking fewer calories than you're burning. So why not monitor your calorie consumption to make sure you're under-eating while at the same time getting enough exercise to burn off a set number of calories? That's your connection.

That was a simple example of how to use mind mapping. However, you can use the same process to tackle more complicated problems. The great thing about mind mapping is that you're not confined to verbal information. If it helps, you can put in images alongside the subtopics, add some colors, and play with the map if you want to keep things intriguing.

Brainwriting

Brainwriting is another variation of the classic brainstorming technique designed to address most of the brainstorming pitfalls discussed earlier in this chapter. Using this method, you can generate a massive amount of ideas while reinforcing connections between them.

The brainwriting process is straightforward: instead of sharing your ideas verbally, write them down and pass them along. There are a few variants of brainwriting. We will first review the advantages of brainstorming and how it could improve conventional brainstorming. Afterward, we will review the most common brainwriting techniques.

Researchers have found that brainwriting helps people be creative without performance anxiety, evaluation apprehension, politics, personalities, and production blocking. Hence, a well-planned and well-managed brainwriting session could be very effective in generating creative ideas. Let's review how brainwriting could be beneficial for creative ideation:

1. Since the participants do not need to speak in a group, they won't suffer from performance anxiety. Public speaking is challenging for many people, making them so anxious that they might hold back their ideas altogether. The group dynamics and human emotions could make this anxiety even worse. Muttered comments, eye-rolling, or sight create a toxic atmosphere in traditional brainstorming sessions. There is a lot less room for such behaviors with written ideas because there is no opportunity for the group to respond to any one written idea.

2. Many working environments suffer from too much internal politics. This could show up in brainstorming sessions in the form of alliances and rivalries between the participants, deteriorating the output quality and damaging morale. Brainwriting helps to control such politics because each individual writes their own ideas privately and anonymously.

3. Brainwriting allows the participants to share their thoughts with a high level of anonymity, as much as totally anonymous in a large group or online brainwriting. This makes it possible for the shyest and the most introverted members of the groups to feel relaxed and express their ideas without the fear of being judged.

4. Production blocking is an essential pitfall of conventional brainstorming. In brainwriting, participants do not need to wait for their turn; everyone can write down their ideas

simultaneously. Hence, brainwriting is especially suitable for large groups for generating many ideas faster. Furthermore, when everyone is busy writing their ideas simultaneously, no one feels "pre-empted" by somebody else's similar idea. Brainwriting even allows people to share their thoughts over some time and without the need for calling a meeting. This could be helpful for team members in different time zones.

5.   Brainwriting is a more flexible process than conventional brainstorming. It could be done in a group or by an individual, all at once or over time, anonymously and without the fear of being judged. In addition, since the ideas are written down, they won't be lost in a conversation and are easier to organize. Also, building on each other's ideas is more manageable because people write their feedbacks. Finally, after a few rounds, the entire pathway an idea has taken is visible, making it easier to make connections and combinations.

Brainstorming has a few variants, all based on the same principles, each with its own approach and suitable for certain situations. Like conventional brainstorming, you always need to set the stage for

Here, we will review the four most common forms of brainwriting:

Basic Brainwriting

After setting the stage and providing the participants with the basic information on the topic, the facilitator presents the question they wish the participants to answer and asks them to share their ideas in writing using small index cards or post-it notes. The facilitator should also explain the ground rules. For example, the amount of time available, the expectations, and limitations, such as the idea should be feasible to implement without spending additional funds.

Participants will have a set amount of time to write their ideas. The available time is usually limited, like a few minutes. This is to prevent the participants from fleshing out all the details or providing justifications for their ideas. Once all ideas are collected, they could be collated by the facilitator. Afterward, the participants analyze the ideas and choose the best ones.

Interactive Brainwriting

This technique has the same principles as those of basic brainwriting. However, instead of collating all the ideas, each participant passes their cards or post-it notes to another team member. Everyone adds their comments or additions on the cards before passing them along. This process continues for as many rounds as desired. The variant of brainwriting encourages the participants to build on each other's ideas, similar to what happens in conventional brainstorming. The only difference is that the interaction process does not include conversation.

Collaborative Brainwriting

Collaborative brainwriting works just like regular brainwriting but more openly and flexibly. Often, this means having a large piece of paper with markers and post-it notes in a public area accessible for all participants. The facilitator writes the question or the problem at the top of the paper and invites the participants to write down their ideas when they have the time of inspiration. These ideas could be both new or built on others' thoughts. The facilitator can set a deadline to collect all the notes, sort, assess, and summarize them.

6-3-5 Brainwriting

This technique works the best with larger groups, which could be divided into groups of six participants. The basic principle is that groups of six people generate three ideas per round, and each round takes around five minutes (hence the 6-3-5 name).

Once the groups are formed, give each participant a printed form with the intended problem at the top and several more boxes underneath the question to add the ideas. The goal is to have three ideas in each round. Hence the exact number of idea boxes depends on the number of ideation rounds.

In round one, each participant writes down three ideas In the second round, the forms are passed along to the next team member, who reads the first three ideas and then jots down another set of three ideas. These could be entirely new ideas or variations of the ideas from the previous round. This process continues in the next rounds until each participant gets their own form back.

If the participants would like to remain anonymous, the facilitator can ask them to put all the papers in a bowl and pick them out one by one. Then, with someone else's paper in your hand, it's time to take a good look at their ideas and do some evaluation. Is it a good one? How could it be improved? Each person should then add three more ideas to the paper that they're holding. This process continues until all idea boxes on each form are completed.

The next step is sorting the ideas by the group members or a decision-maker. The group members could transfer the ideas to post-it notes, cluster them together and refine and combine them before handing them over to a decision-maker. The final step is to choose the direction. This could be done either by the group members or by the management based on the group input.

By following the brainwriting method, participants can build on each other's ideas, issue criticisms, and discuss different concepts, all with some semblance of anonymity.

Although brainwriting has many advantages, it is one of many brainstorming techniques and has its uses and limitations. Brainwriting is the most effective in the following situations:

- Larger groups in conventional brainstorming could be very time-consuming.

- In groups or working environments in which anonymity could help the participants become more open and less anxious, for example, in politically charged groups or groups with various levels of management.

- In larger groups or in online brainstorming sessions in which anonymity is realistically possible.

- Limited time available for brainstorming.

- Groups with dominant personalities that could silence others.

- Relatively straightforward brainstorming topic: for example, "How can we answer customer queries faster?" versus "How can we organize our customer service department?"

- Groups without a trained brainstorming facilitator that can organize, moderate, and guide a verbal brainstorming session.

On the other hand, there are situations in which the chances that brainwriting works well are alow. Here are some examples:

- The group is too small, for example, fewer than six people.

- The brainstorming topic is too complicated and requires close and intensive collaboration between the team members.

- The ideas are too complex to be communicated in a few words.

- The ideas you're asking for are too complex to be communicated in a few words or developed in a few minutes.

- The group members feel uncomfortable about writing.

Brainwalking

Brainwalking is a writing idea exercise similar to brainwriting, but with a key difference: instead of papers moving around, people move around. In other words, instead of people passing along idea cards or notes, they move from station to station, read the ideas, and add their comments or additions. Bryan Mattimore, a specialist in ideation and facilitation, developed brainwalking based as an extension of brainwriting and based on the same principles of cross-pollination and building on each other's ideas. He introduced brainwalking in his 2012 book titled *Idea Stormers*.

Here is how it work: After introducing the problem and analyzing it from various perspectives, the group chooses a few key questions or aspects of the problem as the creative prompt for idea generation. Then, the facilitator writes each prompt on a large piece of paper, such as the A1 flip chart paper, and attaches them to a wall. Next, participants walk between the "ideation stations," They add new ideas or comment on the ones already written on the paper at each station.

Mattimore believes that brainwalking is the "single best technique to begin an ideation session" and mentions several advantages for it compared to other brainstorming techniques. For example, creating more energy in the team, enhancing the sense of shared purpose and group identity, offering some anonymity while openly presenting all ideas, approaching the same problem from various perspectives, the ease of building on other participants' ideas, and so on.

Brainwalking gets people out of their seats, gets them moving, and keeps energy levels up. It is a fun and efficient technique to create a lot of dynamics in the group through its movement and generate many ideas without pitfalls such as production blocking or evaluation apprehension

Reverse Brainstorming

When you get stuck with traditional brainstorming, perhaps it's time to get creative about how to think creatively! One of the most fun and radical brainstorming techniques is called "reverse brainstorming," also known as "negative brainstorming." Reverse brainstorming turns conventional brainstorming upside down. Instead of asking the participants to develop creative ideas to solve a problem or improve a product, you ask them to think of ways to undermine the process, make a goal impossible to achieve, or worsen a product or idea. So, you let all the negative thoughts surface to know what does not work or what can go wrong. Then, once you know what isn't working, you can think of innovative ideas to remove those flaws or shortcomings.

To better understand reverse brainstorming, let's compare it to conventional brainstorming:

- If a business is trying to develop ideas to improve its customer service, conventional brainstorming would ask, "How can we enhance the customer service?". On the other hand, reverse brainstorming will ask, "How can we make our customer service so terrible that all of our customers stop using our products and services?"

- Conventional brainstorming asks, "How can make this project a success?" while reverse brainstorming tries to find ways to ensure the project's failure!

You can use reverse brainstorming on its own as the primary brainstorming technique or complementary to typical brainstorming sessions. The combination could be beneficial in case regular sessions do not deliver satisfactory outcomes due to one or more of common brainstorming pitfalls discussed before or other reasons such as the following:

- Participants have expressed many ideas and have nothing more to add. But there is no clear path yet on how to move forward.

- Encouraging people to think of how something might not work is usually easier than encouraging them to find creative ideas to improve.

- The project is a difficult one, and the participants are tired of it. However, tapping into the frustrations and negative emotions is a great way to make a breakthrough, renew the energy and enthusiasm, and inspire creativity.

- The participants are so connected to the problem and the specific solutions that they cannot think of alternatives. Reverse brainstorming is an excellent technique to help people genuinely think out-of-the-box.

- Participants have focused too much and too soon on an idea and have forgotten other alternatives.

- Participants are trying hard to find creative ideas to improve an existing product or concept. Reverse brainstorming could help to pinpoint the issues and problems. For example, if you want to enhance a travel app, reverse brainstorming could lead to "Try to design a user interface that is very difficult to navigate." This could lead to the discovery that the app is, in fact, impossible to navigate for an average user.

- The group is very familiar with the problem and its challenges and would like to generate many ideas quickly. They know the negatives and what the reverse could look like.

- To empathize with the current or future users of the product or concept and to improve their experiences.

What makes reverse brainstorming unique among all brainstorming techniques is that it can spur creativity and positive problem-solving by tapping into all the negative feelings caused by that problem. In addition, reverse brainstorming offers many other advantages, such as the following:

- For many people, talking about problems rather than trying to find solutions relieves stress and anxiety. Reverse brainstorming leverages this typical tendency positively.

- It can start a journey of discovery as participants begin to highlight the issues and problems, which might otherwise have gone unnoticed.

- Typical brainstorming techniques try to find ideas to make things better. But they might forget to take the flaws out. Reverse brainstorming covers this blind spot of regular brainstorming. In other words, you look for the weak points of a concept or product and how to solve them.

A great example of reverse brainstorming is when corporations and government agencies hire hackers to hack their computer systems and discover their weaknesses. In this "reverse" approach, instead of brainstorming for the best ideas to keep cybercriminals away, you try to mimick those criminals and find out what does not work well in your cybersecurity. Reverse brainstorming is an excellent tool because it teaches you what not to do. In that sense, try to use reverse brainstorming often, either as the sole brainstorming technique or after trying other methods.

Starbursting

Starbursting creates a map of questions using a six-sided star with the core ideation topic at the center. Each side of the star contains one of the following fundamental question words: who, what, where, why, when, and how. It's a straightforward and yet powerful method proposed ideas or finding new ones and making connections between them.

Starbursting sessions focus on the nuts and bolts of the ideas gathered previously. Hence, starbursting is beneficial in conducting the convergent phase of the brainstorming process or extending the divergent phase. It starts with simple and obvious questions and proceeds to more complex and impactful ones. The goal is to understand and challenge the ideas and ensure that all aspects of the selected ideas have been considered before choosing the final pathway.

Starbursting is similar to questionstorming but has some significant differences too. Unlike questionforming, starbursting allows the participants to answer the questions immediately and discuss the answers. In addition, starbursting is more systematic in the type and flow of the questions and analyzing the responses.

Imagine a business is trying to boost its sale. The outcome of previous brainstorming sessions was developing and launching a new mobile app to make shopping easier for mobile users. The core question for the starbursting session is how to create and launch the mobile app. After writing this question at the star's center, the team thinks of multiple questions using each question word. Here are some examples:

- **Who** is the intended audience for this app?

- **Who** are the typical first users?

- **Who** are our competitors?

- **What** should we call it?

- **What** are the desired functionalities of the app?

- **What** are the unique selling points for this app?

- **Where** are we going to make the app?

- **Where** are we going to advertise it?

- **Where** are most of the customers located?

- **When** do we want to launch the app?

- **When** are we going to advertise the app?

- **When** do we expect to see the impact on our sales?

- **Why** is now a good time to launch this specific app?
- **Why** will the customers like this app?
- **Why** do we think it's better than the competitor apps?

- **How** will we organize the development team to achieve our goals with this app?
- **How** will we conduct the market research to define the customers' needs and preferences?
- **How** will we integrate our current services to the new app?

Once there are at least three questions on each corner of the star, the team systematically discusses the questions and jots down a short answer for each one.

Starbursting is an excellent technique to generate focused questions to explore, analyze, and refine a proposed idea or product. It challenges the assumptions and fixations on a particular concept. On the other hand, starbursting is less effective as the primary brainstorming tool since it requires a starter idea or topic.

The biggest pitfall of starbursting is that it's an open-ended process, which could make it tricky to keep the focus and have a tangible outcome within a reasonable time. Hence, proper facilitation is critical to managing the questions, the participants, and the time.

For instance, in the starburst session for the new mobile app, John might look at the "Why" corner of the star and ask: "Why don't we hire new staff to build and launch the app instead of loading the exiting staff with even more work?" John might have a good point and has asked a question starting with "why". But such a question will distract the team from the topic under discussion take them to the treacherous waters of human resources and corporate finance.

The facilitator has a vital role in handling off-track questions. Addressing them will de-rail the discussions, and ignoring them could result in non-cooperation. There are two main options to take such questions safely. The first one is to create a separate list of off-topic but significant questions and discuss them at another time. The second option is to ask the team members to rephrase the question in the context of the discussion topic. For example, John's question could be restated as "How could our development staff could complete this project without working overtime?"

Starbursting can suffer from some common brainstorming pitfalls discussed earlier in this chapter, such as evaluation apprehension and dominant personalities. Therefore, the facilitator should make sure that all participants can share their ideas. The standard measures such as using a timer to prevent anyone trying to "filibuster" the session, encouraging all participants to come up with questions, handle off-track questions, etc., will all help. Furthermore, the facilitator could give the participants some time to think up their questions first and discuss them afterward. This is like adding some brainwriting to the starbursting session.

Managing time is the third challenge of the starbursting facilitator. Production blocking and scope creep combined with the open-ended nature of starbursting could easily kill the session. However, proper facilitation could resolve such issues to a great extent. Here are some strategies to manage the time in a starbursting session:

- Allotting a limited time for each portion of the process. For example, five minutes to think up the questions for each corner of the star, three minutes to discuss each question, etc.

- Limiting the number of questions

- Assigning the discussion of each set of questions to a working group: This could work well in larger starbursting groups.

- Using brainwriting or brainwalking to discuss the questions faster.

Rolestorming

This particular method is supposed to make you feel more comfortable in suggesting wildly innovative ideas. It addresses the fear of criticism by allowing team members to take on a "role" and present ideas as that character. It's a lot like role-playing but used to benefit the business or any other situation where you find brainstorming useful. Rolestorming a very useful and playful technique to generate new ideas.

Rolestorming is a unique type of group brainstorming that involves role-playing. For example, participants could play real-world roles, such as a customer, competitor, manager, or fictional characters like superheroes or celebrities.

The premise of rolestorming is that brainstorming while playing a role increases the chances of shedding pre-suppositions and thinking in new and creative ways. Introduced by the business guru Rick Griggs in the 1980s, rolestorming is a group brainstorming method that creatively addresses the most common brainstorming pitfalls. During his career as a manager in various Silicon Valley corporations, Griggs noticed that many brainstorming sessions suffer from the following pitfalls:

- The first idea is the best.
- Evaluation apprehension, i.e., the fear of looking stupid,
- Following the dominant personalities,
- Difficulty with on-the-spot creative thinking.

The main reason rolestorming could work well is reduced participant anxiety. It seems like most people feel more comfortable expressing an idea when they are in character. In addition, they can think more creatively because playing a role allows them to see the problem from another perspective and get less fixated on the details.

The type of personas for participants greatly influences rolestorming sessions. A difficult or demanding customer, a board member, a historical figure related to the topic of the session or famous for a particular type of thinking, a superhero or a supervillain, and countless other options are possible personas. The participants could all take the same type of persona or different types. They can also choose their character or get a character assigned to them.

It is important to remember that rolestorming is more like a game and less like many other brainstorming processes. Hence, the rule of rolestorming focuses more on having fruitful and creative improvisational acting. Here are some examples:

- The participants should decide or know their characters and could describe their qualities.

- Everyone should immerse in their characters and avoid referring to actual circumstances or limitations such as lack of funding or staff.

The participants need to remember that rolestorming is more like a game, and they should be willing and prepared to play along. The whole purpose of this type of brainstorming is to get past the daily struggles, set the mind free and approach the problem from angles that might otherwise remain unexplored.

Because of the unorthodox nature of rolestorming, the role of the facilitator is even more critical than in other brainstorming techniques. The facilitator of a rolestorming session should preferably have some acting or improvisation experience and should be comfortable with unpredictability and even silliness. In addition, they should be good at note-taking, concluding, and establishing connections and have a keen eye for clues and potential breakthrough ideas. For example, if a participant in the character of an unhappy customer complains that " The lines are always busy!" the facilitator should jot that down as a potential lead.

Some people love rolestorming and feel very comfortable performing it. But many people find it confusing and even embarrassing. It is the facilitator's responsibility to explain the game and its rules to the participants and help them feel relaxed and comfortable to play along.

A great way to kick-start a rolestorming session is to use some warm-up questions or icebreakers. These can be fun and interactive questions or games related or unrelated to the topic under discussion. For example, ask them, "If our company was an animal, what kind of animal would it be?!

Finally, it is crucial to know how to organize and use the ideas generated in the session. Once more, the facilitator has a vital role. They should be able to take the plays and improvisations and turn them into viable breakthrough ideas. Then, if needed, the participants could have a follow-up session to examine those ideas and help the facilitator to decide on the path forward.

Braindumping

People rarely start a brainstorming session without their preferred ideas and approaches. While it's good to know the problem prior to the brainstorming session, entering the sessions with your head full of ideas could be disastrous for your creative thinking.

Dee Hock, the founder and former CEO of the Visa credit card association, once said that *"The problem is never how to get new innovative thought into your mind, but how to get old ones out."* Any facilitator or participant of a brainstorming session should be wary of their existing ideas because most of us try to advocate our pre-destined ideas, regardless of what brainstorming technique is in use. Such a tendency to cling to our ideas also makes us less open to other people's thoughts.

Braindumping is an effective technique to avoid the drawbacks mentioned above. It is a very simple exercise done at the beginning of a brainstorming session to help the participants break free of their pre-determined ideas. With a braindump, everyone writes down their existing thoughts and anything that pops up in their mind on the topic. They "dump" their ideas in the first few minutes of the session to empty their minds and get ready to find new and creative ideas. Then all ideas are shared with the group and briefly discussed but not criticized. The facilitator should explain that these ideas are recorded and will be part of the end result. Once the participants know that their beloved ideas are safe, they are more inclined towards thinking up alternatives. Now, it's time to continue the session with other brainstorming techniques and look for ideas with a fresh perspective.

It is fair to say that every brainstorming session should start with a braindump. Spending 10-15 minutes at the start of the session to set the participants free of their pre-destined thoughts will invariably enhance the chances of having a fruitful creative thinking session.

Brainsteering

This is a modernized version of the original Osborn process. It follows the same principles but takes the company culture and targets into account. In that sense, it could be a great corporate brainstorming tool.

Introduced by former McKinsey consultants Kevin and Shawn Coyne in their 2011 book *Brainsteering: A Better Approach to Breakthrough Ideas*, brainsteering takes the conventional brainstorming and "steers" it more productively and practically, especially in corporate environments.

Attempts at brainstorming new ideas in corporations often fail due to the common brainstorming pitfalls such as evaluation apprehension and ignoring the organization's targets, culture, and limitations. Instead of simply thinking of as many ideas as possible, brainsteering focuses on delivering practical and creative ideas within the boundaries of organizational targets, norms, culture, and constraints.

Brainsteering follows the rules and principles of conventional brainstorming, i.e., the initial focus on quantity of ideas, deferring judgments, divergent-convergent thinking cycles, combining and refining ideas, etc. However, what makes brainsteering novel is putting the entire brainstorming process in the context of a very carefully defined box. So, instead of presenting a broad question and asking the participants to think outside the box, brainsteering first represents a box, determined by what the corporation wants or is willing to consider, and then encourages the participants to think creatively within the boundaries of that box. Once the participants know what direction to channel their mental energies, the chances of having a more productive session are higher.

In principle, brainsteering is brainstorming while taking real-world circumstances into account. The users of brainsteering are not only corporations. Universities, non-profit organizations, and individuals can also benefit from this technique. For example, the Coyne brothers have helped many universities to apply brainsteering for cost reduction projects. In one example, the goal was to reduce the cost and footprint of the computer systems. The university found out that nowadays, most students use their personal laptops for web browsing, doing homework, etc. So, the university decided to subsidize buying a new laptop for the students and eliminate all on-campus computer sites.

As mentioned before, brainsteering follows the basic rules and principles of conventional brainstorming. However, there are specific guidelines to have successful brainsteering sessions. Let's take a closer look at the seven golden guidelines for brainsteering:

1. *Know your targets and restrictions.*

If the brainstormed ideas are beyond what the organization is willing to consider, those ideas won't do that much. Thinking out-of-the-box will be just a buzzword if real-world circumstances or company policies require you to live within certain boxes.

Brainsteering starts with understanding the organizations' targets, decision-making process, restrictions, and limitations. These will define the box of viable ideas. For example, imagine a bank is brainstorming on how to reduce the operational costs, and the outcome is "update the IT systems." However, if the senior management has already fixed the IT agenda for the next two years, the brainstorming session was a waste of time.

It would have been much more productive if the brainstorming planner collaborated with the senior management to define a common target tailored to the organizational needs and circumstances. For instance, good ideas would require no more than a certain budget and could generate profits or savings in specific areas in line with the general organizational plans. With a carefully defined box, the participants of the brainsteering session would not waste their time and energy on any idea requiring regulatory approval, big budgets, organizational change, etc.

The same premise applies to brainsteering by individuals. For example, you want to have an additional source of income but cannot take on a 2nd job or work overtime because you have to take care of your young family. For your situation, maybe an online side hustle is a more viable option since you might be able to work on it in the evenings, weekends, or any other possible time.

2.   *Ask the right questions.*

A misconception about brainstorming is that brainstorming lets the mind roam free, hoping that inspiration and brilliant ideas will follow. Unfortunately, in practice, this approach rarely works. Instead, successful brainstorming needs preparation and facilitation, most importantly, a clearly defined question (or questions). Building the ideation session around a few well-defined questions is the basis for some of the best variants of conventional brainstorming, such as questionstorming and starbursting.

Good questions have the following two characteristics:

- *They require the participants to follow a new approach*: Forcing the participants to distance themselves from their existing thinking patterns (i.e., what worked in the past) and look at the problem from a new perspective.

- *They are specific enough to define clear boundaries for the ideation space*: To steer the ideation efforts more productively without being too restrictive.

For example, a consumer products manufacturer trying to develop a new product could start the brainsteering session with questions such as "What's the biggest avoidable hassle our customers have?" or "How does a customer might use our products in unexpected ways?"

The number of questions depends on the size of the brainsteering group, the problem at hand, the ideation scope, etc. However, there should be at least a few questions to explore various aspects of the problem.

3. *Choose the right participants.*

Another common misconception about brainstorming is that involving people without knowledge or background on the discussion topic is a good approach because they can bring in new perspectives. However, as we discussed earlier in this book, creativity, like expertise, is domain-specific and requires a certain level of knowledge on the intended topic. Simply put, for any brainstorming session to be productive, it is essential to involve the people who can answer the questions. In brainsteering, this is a basic rule: the participants should have first-hand "in the trenches" knowledge on the discussion topic.

Coyne brothers give a great example of this in their book *Brainsteering*. They talk about one of their retail customers brainsteering to reduce the rising percentage of debt delinquency rate. When one of the participants asked, "What's changed in our operations which could have caused this increase?" a frontline manager replied, "Death has become the new bankruptcy!" Upon further discussions, it became clear that some customers who fell behind their payments falsely claimed bankruptcy or even instructed their household members to claim they're dead, knowing that the collection agents will stop pressing the issue.

The presence of a line manager with specific knowledge of the issue created the opportunity to determine the root cause of the problem. After some discussions, the team concluded to instruct the collection agents to firmly but sensitively ask for more details if they suspected a ruse. Dishonest customers would invariably reveal themselves if asked for more information, and the collection process could continue.

4.   *Set the stage.*

Participants of a brainsteering session should receive proper orientation at the beginning of the session. The facilitator should explain the expectations, the targets, and the constraints. In addition, they need to explain the brainsteering process and the differences with traditional brainstorming.

Brainsteering is usually slower but deeper than brainstorming. The majority of the discussions occur in smaller subgroups, as we'll review in the following guideline. It will help if the facilitator gives examples of past brainsteering sessions and shares some success stories to inspire and motivate the participants.

5. *Form subgroups to discuss individual questions.*

Instead of having one long, continuous discussion among the entire group, divide the participants into subgroups of three to five people and instruct them to conduct multiple highly focused ideation sessions. The smaller subgroups encourage the participants to speak up, whereas the norm in bigger groups is to stay silent. Assign a few questions to each subgroup and ask them to discuss the questions for a fixed period, for example, half an hour.

Proper facilitation of the subgroups is crucial, especially handling the participants that prevent others from sharing their ideas, for instance, the people with higher organizational ranks (the bosses) and those with big mouths. The facilitator can also instruct the subgroups to use techniques such as braindumping and reverse brainstorming.

Some of the guidelines mentioned above do not apply to individual brainsteering. When you are trying to find creative ideas to solve a problem within your means and possibilities, there are no subgroups, and probably you are thinking alone.

The key is to focus on the core principles of brainsteering, that you should have a clear picture of your goals, possibilities, and limitations and try to think within those boundaries. You can always get advice from other people. In that case, you will be both a participant and the facilitator of the ideation process.

6.   *Wrap the session up.*

Once all subgroup discussions are over, the worst thing to do is let the entire group choose their favorite ideas from the pile. There are two reasons for this:

- The participants might lack the executive-level understanding of the priorities and the decision-making criteria to choose the most viable ideas.

- Choosing a winner idea in front of the group could be a demotivating experience for many participants. It could even be irrelevant if the real decision-makers overrule what the brainsteering group chooses.

A better approach is to ask each sub-group to narrow down their ideas and choose their favorites, then share them with the whole group with the note on the next steps and when to expect the winner ideas.

7.   *Follow up as soon as possible.*

A brainsteering session needs two sets of follow-ups. The first one is with the senior management for decision-making, and the second is with the participants to share the outcomes.

The follow-up with the management should be thorough and without delay. The senior management could evaluate the ideas for immediate implementation, immediate rejection, implementation at the closest appropriate time, or further evaluations. The facilitator or the management should inform the participants about the decisions and the underlying reasons.

Stepladder Brainstorming

As the name implies, this is a step-by-step process to create a "ladder" of ideas to solve a problem or make a decision. The primary advantage of the stepladder technique is that it encourages everyone to participate, even those who tend to be shy and are quiet. It starts by presenting the problem to everyone in the group. Once everyone knows of the problem, let everyone leave the room except for two people. Next, let those two people collaborate to come up with ideas. Once they're done, add one person into the room. The new person gets to share his thoughts before the first two share theirs. Afterward, let someone else walk into the room and repeat the process until all the group members are in the same room. Once everyone has shared their ideas and all the team members are in the room, you can start the discussions to reach a concluding decision.

The stepladder technique was initially developed as a decision-making tool. The premise behind it is that in every team, some members are more aggressive in expressing their opinions. Therefore, to hear everyone's opinions, it is essential to give all members a chance to speak up their minds. That's why As we discussed earlier in this chapter, dominant personalities could highjack brainstorming sessions. Hence, the stepladder could be an excellent brainstorming approach if one or two team members are more outspoken than others or when some participants hide among the group.

Charrette Procedure

The Charrette procedure is designed to help a large group of people to think about one or more problems. The name comes from the French word "charrette," meaning "cart."In the 19th century, a cart was used to collect and transport the drawings of student architecture for marking. The Charrette process does the same thing.

First, the participants are divided into smaller groups of, for instance, five people. Then, all the subgroups start to brainstorm about the problem simultaneously. The ideas from each group are handed over to the next group for discussion and refinement. Finally, all ideas are gathered, discussed, and prioritized. If there are multiple discussion topics, each subgroup will start with one subject. Then the subjects and the ideas move from group to group.

The Charrette process is an excellent brainstorming technique in the following cases:

- There are a large number of participants (e.g., more than fifteen).

- There are multiple discussion topics

- The available time is limited, or the organizers would like to limit the discussion time deliberately.

- The organizers would like all the members to participate in the discussions.

The Charrette procedure could produce high-quality outputs because each subgroup polishes and refines the most popular ideas.

Online brainstorming (aka Brain-netting)

So far, we've talked about idea generation in face-to-face groups. What if the participants work remotely, in different countries or multiple time zones? That's where online brainstorming comes in handy!

Online brainstorming is conducting a brainstorming session online and computer-mediated. With the new tools and software, online brainstorming is more effective now than ever before. Furthermore, with the recent remote working trend, online sessions are becoming a norm and sometimes the only viable option in many organizations. Hence, it is worthwhile to discuss online brainstorming and its tools in more detail.

Online brainstorming is not a distinctive technique such as rolestorming or brainsteering. It is more about conducting the existing methods effectively and productively using a wide range of digital tools. Nowadays, almost everyone uses group calls, chat apps, and video-conferencing as the primary ways to communicate online (apart from emailing each other!). But what if the team members want to actually work together on shared documents or whiteboards, communicate like a face-to-face brainstorming session, and collaborate in activities such as mind-mapping?

A simple and accessible option to allow team members to collaborate online is using collaborative documents such as Google Docs, which would enable multiple users to write, edit, and mark up a document or a spreadsheet and add their comments. However, if you like to talk while editing, you need to combine Google Docs with a conferencing system like Skype.

The next level is the all-in-one business suites which allow the team members to talk, share screens, access shared folders, have video conferences, etc. The most common options are Skype for Business, WebEx, and Microsoft Teams.

When it comes to brainstorming, online tools could be priceless, especially if some or all of the participants are working remotely. There are tons of apps with different features and price ranges to help you and your team with online brainstorming. These online tools are tailor-made for collaborative digital brainstorming and offer functionalities not available in standard business suites.

There are two main types of online brainstorming tools: online mind mapping tools and virtual boards. Here are some examples of each category:

Online mind mapping tools

1. *Bubbl.us*: This web-based platform creates mind maps and allows you to build a tree of ideas by starting at a core idea and then adding new ideas at different levels. You don't have to download any app; everything is web-based. Bubble.us comes with three plans. The basic functions are free, but you can use the paid version after the 30-day free trial for better features.

2. *Freeplane*: Freeplane allows you can build and connect topics and subtopics and insert colors, shapes, and images as needed. This tool can also help you classify, position, and order different factors related to the problem. Moreover, it's very user-friendly since you can drag and drop items on the screen.

3. *Popplet*: Popplet is a good choice if you're trying to encourage kids to brainstorm as part of school activity. It also works for office situations because the app lets multiple people access it at the same time. In addition, you can build presentations with visualizations and diagrams. The app is primarily developed for Android, but it's also available for iOS users.

4. *MindMap*: Available via Google Chrome's extension, the MindMap is another excellent brainstorming tool, with Google Drive, Dropbox, and Cloud all built-in as part of the

support system. Using this tool, you can even hand-draw an image and attach it to your mind map.

5. *FreeMind*: For individual brainstorming, FreeMind is a good option. You can use it on Windows, Linux, or Mac computers. It has a one-click navigation panel and lets you upload images, drag-and-drop, and even import any existing mind maps.

6. *Coggle*: This tool helps you to create interactive mind maps in the form of intricate visual networks. The team members access a shared min map, brainstorm ideas on the core statement in real-time, and add it to the mind map. The map has several branches shown in different colors. Each branch contains a set of related ideas.

7. *MindMeister*: This is a powerful web-based mind mapping tool with collaboration features such as an integrated chat function and vote and comment on the ideas. The service is free, working on up to three mind maps. The paid version is still very affordable at $ 4.99 per month for unlimited mind maps.

Virtual boards

1. *IdeaBoardz*: This is a free web-based virtual board where collaborators can add their ideas using virtual sticky notes,

organize ideas in different sections, vote for their favorite ideas, sort ideas by the highest number of votes or sections, search keywords, and export the boards for later discussions.

IdeaBoardz provides the digital version of a flip chart or whiteboard and could be used with a wide range of brainstorming methods such as online brainwriting or reverse brainstorming.

2.   *Realtime Board*: This is a collaborative, online whiteboard space with virtual post-it notes. In addition to adding their ideas on sticky notes, the team members could add files, images, and documents. Once the session is complete, the notes and other files could be exported into a presentation or a PDF file. Realtime Board has both free and paid versions.

3.   *MURAL*: This web-based visual collaboration tool offers a virtual board, powerful facilitation features such as timer and voting, and a digital canvas to add shapes and notes.

Tools such as the ones mentioned above are essential to conducting an online brainstorming session. However, like any other brainstorming technique, proper planning and facilitation of the session are essential. Furthermore, online brainstorming has its unique challenges. For example, collaborating in groups across time and space poses some fundamental issues uncommon in face-to-face brainstorming. Therefore, to make online brainstorming a productive endeavor, the facilitator and the participants should be aware of specific issues and follow strategies to handle them. Let's review some of these issues and the rules of thumb to minimize their impact:

- Having team members in different time zones could result in serious scheduling issues. After all, finding the right time to have people from Singapore, London, and San Fransico in the same online meeting is not easy. In many cases, at least some parts of the online brainstorming have to happen asynchronously. We will elaborate on this shortly.

- Not all people are technically savvy. Some participants might not be able to use the online brainstorming tools sufficiently or to troubleshoot IT issues.

- If some participants use Mac computers and others PCs, there might be some compatibility issues.

- Efficient use of online brainstorming tools takes time and practice. The first few sessions could be messy and unproductive because the participants are still learning.

- If the team includes external contractors or freelancers, they might not have access to your internal network or the paid version of the brainstorming tools.

It is possible to solve all of these issues relatively easily, except the first one. People soon learn to use virtual boards and mind maps and solve IT issues. But collaborating over different time zones is more challenging to resolve. No matter how sophisticated the tools are, eventually, brainstorming requires the collaboration of a group of human beings. How can a group work together if some participants are sleeping while others are in the middle or at the end of their workday?

The solution is to break up the brainstorming process into asynchronous and synchronous parts. Online audio or video calls are the "synchronous" times of the collaboration, meaning they require everyone to interact simultaneously, as opposed to "asynchronous" times of the process, which are times when people can complete parts of the process on their own.

A good approach is to send the intended question, plus some context, the goals, and constraints, to the participants by email or using a shared document such as a Google Doc and ask the team to think about the topic at whatever time is best for each team member. In this way, you use the asynchronous part of the brainstorming process to think up ideas. Then, team members can share their ideas with the facilitator or directly add them to the virtual board or the online mind map.

The next step will be to have the entire team together in a voice or video call to discuss the ideas and think about the topic in new ways. This is the synchronous part of the process to share, discuss, and combine ideas.

This asynchronous-synchronous cycle can happen several times until reaching the desired outcome. For example, after the first call, participants can take some time to think about the ideas and what they heard on the call, add new ideas to the board or the map, or sort and combine the existing ones. Then, they can come to a second call with their feedbacks.

We talked about interactive brainwriting earlier in this chapter and how people can share their ideas and interact over time in asynchronous and synchronous parts. The opportunity to take the time and think about proposed ideas can lead to more thoughtful brainstorming. In that sense, online brainstorming offers the benefits of brainwriting.

Brain-netting, just like conventional brainstorming, requires skillful facilitation and understanding of the problem, process, participants, goals, and constraints. Furthermore, the organizers of an online brainstorming session should select the right tools and methods and manage asynchronous collaboration. Some of the other brainstorming techniques, such as braindumping, could be incorporated into online brainstorming. Finally, don't forget to follow up with the participant about the outcome and experiences from the session.

We covered a wide range of brainstorming techniques in this chapter. The following two chapters will discuss applying these techniques in group and individual brainstorming in professional and personal contexts. Productive brainstorming can feel like a win for everyone as if the process was worth their time and effort. So, how should you choose between all those tools and apply them to generate creative ideas in your business or personal life? We'll find the answer sin the upcoming chapters!

Chapter 5: Group Brainstorming

Group brainstorming lies at the heart of collaborative idea generation in many professional environments all over the world. If done correctly, it can encourage creative thinking, strengthen the team spirit, and help the team land on breakthrough ideas. We have already discussed the origins and principles of brainstorming and the related techniques in the previous chapters. This chapter will focus on how to have a thriving group brainstorming session, face-to-face or online. In addition, we will address effective facilitation, brainstorming mistakes to avoid, and how to leverage the digital tool for better brainstorming.

Is Group Brainstorming Still Relevant?

Before we dive deep into the nitty-gritty of group brainstorming, let's take a moment first and see whether the idea of group brainstorming still makes sense. If you tell a colleague at work that your next meeting is a brainstorming session, the chances are high that you will get eye rolls and groans. For many people, brainstorming is a concept that makes sense but does not work in practice. Even worse, many consider brainstorming outdated and a waste of time. So, why is brainstorming still being used by the corporate world? Is it still a relevant and valuable concept?

Let's keep the comparison between the group and individual brainstorming for the next chapter and first focus on why group brainstorming has a bad reputation with some people. Most critics of brainstorming and those with negative brainstorming experiences think of traditional brainstorming only and not the more modern variants discussed in the previous chapter. Here are some of the most common complaints about group brainstorming:

- Voicing ideas in a group setting is uncomfortable for many, for example, because of the fear of being judged.

- Not everyone is good at rapid idea generation.

- It takes too long.

- Not everyone participates in the thinking process.

- The group can get fixated on a particular idea.

- You have to wait for others.

- You end up with obvious ideas.

- The participants tend to favor more conformist views.

- Introverts are less likely to vocalize their ideas.

- One person talks, the rest have to listen instead of thinking of their own ideas.

We discussed all of these complaints as the pitfalls of traditional brainstorming in the previous chapter. Furthermore, we saw how the new brainstorming techniques address almost all of the typical brainstorming drawbacks while still benefiting from the group synergy effect. For example, brainwriting can help introverts share their ideas without voicing them in the group, or braindumping takes the obvious ideas away. Similarly, the stepladder or Charrette encourages everyone to participate in the thinking process, and brainsteering provides a pragmatic approach for productive brainstorming in personal and professional set-ups.

The potential benefits of group brainstorming are enormous. The most significant one is the group synergy, i.e., taking advantage of the group's collective experience and mental power. The group synergy allows the members to create something bigger than the sum of its parts. When one member is stuck with an idea, another member can use their creativity and knowledge and take the concept to the next level. Similarly, the team members can give feedback, build on each other's ideas, and combine and refine them. Furthermore, successful group brainstorming promotes the group ownership of the new ideas because everyone contributes to creating them.

But does group brainstorming always work? Not necessarily. Is it possible to conduct a productive group brainstorming? Absolutely! Performing group brainstorming is like doing a DIY project. If you have the right skills, use the right tools, plan and budget the project wisely, you can get the DIY project done with acceptable quality, potentially saving a lot of money.

Similarly, successful brainstorming needs the right tools and techniques and proper planning and execution. Saying that "Brainstorming doesn't work!" or "Brainstorming is outdated and a waste of time!" is as valid and relevant as saying "DIY projects never succeed!" Sure, many people regret some of their DIY projects. But countless other people plan and complete many DIY projects skillfully. So, the real answer to the question "Does group brainstorming work?" is "It depends!"

Productive group brainstorming comes down to two main factors:

1. Choosing the most suitable brainstorming technique.

2. Proper set-up and execution, before, during, and after the session.

So, what key steps should you, as a participant, facilitator, or organizer, take to make the most out of your group brainstorming sessions? What are the do's and the don't's for successful group brainstorming?

Chapter 4 discussed the brainstorming tools and techniques in detail. There is always a suitable technique, whether face-to-face or online, with a small or a big group, or for group or individual brainstorming. You can always revisit Chapter 4 and choose the technique that fits your context the best. As you can see, I am not referring to Chapter 2. In other words, I do not recommend using the traditional Osborn-style brainstorming to avoid the common pitfalls in the first place.

This chapter discusses the second factor for productive group brainstorming, i.e., the session's set-up, planning, and execution, especially effective facilitation and management of the group dynamics. We will start by setting the stage for the brainstorming session and then continue with every phase of the process step by step.

Perhaps you will be a participant in most of your brainstorming sessions, but you might become the facilitator or the organizer from time to time. The best way to learn the ins and outs of the process is to imagine yourself as the facilitator of your intended session. The facilitator is involved in the planning and set up of the session, pays attention to the group dynamics and personalities, actively steers the session, and can change its course of action. Knowing the principles and practices of an engaging brainstorming session will also undoubtedly benefit you as a participant. In fact, the ideal scenario is if the participants are as knowledgeable on the process as the facilitator. So, in the rest of this chapter, you are the facilitator and the organizer of an imaginary brainstorming session, trying to have the most fantastic brainstorming ever!

Ten Surefire Ways to Kill a Brainstorming Session

Let's do with some negative brainstorming! How can you have an awful brainstorming session? A session that frustrates and demotivates everyone; organizers, participants, facilitators, managers, etc.? What defines such a session is what you should avoid in your brainstorming session.

In my experience, there are ten primary characteristics of a dull and unproductive brainstorm that sucks up time and energy, frustrates everyone, and delivers nothing. If you can manage to avoid these ten follies, you will be ahead of the vast majority of brainstormers.

1. *Killing the real purpose of brainstorming*

Let's start with the basics. The whole point of brainstorming is to let creativity emerge and shine. If you want to ensure your session goes nowhere, stifle creative thinking! There are countless ways to do that. For example, you can try saying the following anywhere in the session:

- "Everyone, start tossing out ideas!"

- "We need to walk out of here with one great idea!"

- "Don't waste our time with nonsense!"

- "Let me kickstart the session by telling my ideas!"

- "We've tried that before!"

- "Everybody, please stay seated!"

    Each of these statements signals the wrong mindset and poor facilitation and can sabotage the whole meeting.

A call for action like "Start tossing out ideas!" inspires nobody. Creative thinking or problem-solving is a process. It takes time and happens via divergent and convergent thinking cycles. Pushing people is an awful way to encourage creativity. The outcome is usually that the usual suspects take over the meeting while the rest are doodling or daydreaming.

What if the session generates more than one good idea? Or maybe no clear breakthrough idea? Focusing on a pre-determined outcome chokeholds the creative thinking process because the participants will focus on satisfying the facilitator or the organizer. Instead, the focus should be on engaging the participants, having a nice flow of ideas, and creating positive and productive group dynamics.

In many cases, participants of the group brainstorming session are not even the decision-makers to choose one great idea. A good brainstorming session has a well-defined problem statement and clear goals and constraints. But that does not mean setting rigid goals. If the session is well-planned and well-executed, there is usually no shortage of good ideas.

Statements like "Let's walk out of here with one great idea!" and "Don't waste our time with nonsense!" usually go hand in hand. When the facilitator or the organizer of a brainstorming session says these words, they are determined to save time and energy and guarantee a bunch of humdrum ideas! How can we know an idea is nonsense? If it does not make sense at first glance, is it nonsense? It is not uncommon to find out that many ideas are not viable or not advantageous. But to reject them fast and furious goes against the whole purpose of brainstorming.

Brainstorming should be a safe haven for creative thinking, exploring new ideas, and experimenting with new concepts. It should allow the mind to be free and playful, or at least not labeling any idea as "nonsense." All breakthrough ideas might look ridiculous at first. One of the worst things you can do in a brainstorming session is to disallow "nonsense"! On the contrary, the organizers of facilitators should encourage sharing wild and crazy ideas. Maybe they prove to be insane or impractical. But they can provide the breeding ground for sane and sensational ideas.

Asking people to remain seated or even implying that is another common mistake. Movement creates energy, which results in more and better ideas. The brainstorming session should have a relaxed and informal atmosphere to remove anxieties and set the minds free. It should be fine to talk while standing or walking. In fact, several brainstorming techniques such as brainwalking, the stepladder, or collaborative brainwriting heavily rely on creating movement in the group.

Finally, we come to the deadliest statement: "We've tried it before!" and its many variants such as "We don't do it that way" or "The world doesn't work that way!" etc. These statements are like killing an idea with a shotgun and are totally counter-productive for creative thinking. The point here is to avoid such messages in the divergent phase of the brainstorming process. Of course, analyzing and editing are essential parts of successful brainstorming. But only when all ideas are gathered and it is time for the convergent phase.

Ideas are fragile and wrong words bring the whole session to a halt. So if you want to have an intriguing and engaging brainstorming session, steer clear of statements such as those mentioned above.

2. *Being ill-prepared*

Poorly planned and prepared brainstorming session amounts to nothing more than wasted time, frustrated participants and a stack of half-baked ideas.

Preparation for a brainstorming session has many facets and includes deciding on various factors. Here are some examples:

- Basic parameters of the meeting, such as the size of the group, the duration of the session, the participants, etc.

- Brainstorming technique and format (in person or virtual)

- Communication with the stakeholders

- Choosing a skilled facilitator (in case you are the organizer)

Maybe the brainstorm session will take only 30 minutes. But work should begin well before the team gets together. First, the person who leads the session should clearly define the question and its context and constraints. Then, once the basics of the session and its scope are set, the facilitator can prepare some guided questions to put the team in the right direction.

When we speak of preparation, we're also talking about having the people armed beforehand, giving them the right information prior to the brainstorming session. Then, you can ask them to think up their ideas and use the session to share and discuss them.

The great thing about sharing the discussion topic and the background info in advance is that you allow the ideas to mellow in their head, creating an incubation period. This helps inspire other ideas during the session. Even better, the initial understanding of the challenge means you don't have to waste too much time with the explanation.

Taking all these steps takes time and energy and goes against what many people think of brainstorming. A common misconception is that brainstorming is cobbling a meeting on short notice, inviting a few folks, and attending the session with little preparation. Many think brainstorming is a spontaneous process and better happen in the spur of the moment. They might even think preparation and facilitation could hinder the free-flowing generation of ideas.

The notion that you have to let go of the structure to be creative is a fundamental misunderstanding of the brainstorming process, especially group brainstorming. Like any other group activity, planning, preparation, and guidance are critical for success. Letting go of the structure could be beneficial in the divergent brainstorming phase but not in the initial or final stages. We will discuss the preparation steps in more detail later in this chapter.

3.   *Unclear or wrong question*

Sometimes the participants go through a brainstorming session to find out that they have not addressed the core problem because the problem statement was incorrect, inaccurate, or poorly phrased. The same problem rephrased in different forms will have other answers. For instance, suppose you (or your department) have a tough time coping with the workload. Looking for ideas on how to handle the situation, you might ask yourself, "How can I finish all my work before the month ends?" The possible answers would be working longer hours, working on the weekend, working more efficiently, etc. But imagine you replace the question with "How can I get all the work done before the month ends?" This question approaches the problem from an entirely different perspective and leads you to ideas to involve other people to do the work together.

Perhaps a real-world example will make the point clearer. In 1955 the first commercial television started broadcasting in The UK. The broadcasting rights were auctioned in 1954, with many parties trying to determine which region will generate the highest advertising revenue. Most analysts assumed that wealthy regions will be the biggest spenders. Hence, they asked themselves, *"How do we get the broadcasting rights for the wealthiest regions?"* But One man had a different approach. Sidney Bernstein from Granada Television argued that instead of bidding on the wealthiest regions, he should focus on the highest watch time areas. So, he asked himself, *"How do I get the broadcasting rights of the regions where people watch TV the most?"* His different take on the problem led him to bid on the wettest regions, where people spend the most time watching TV! As a result, Granada Television eventually got broadcasting rights in the rainy north of the UK and became one of the most successful British TV production companies.

When trying to formulate the brainstorming question, it is usually easier to focus on the symptoms rather than the root cause of a problem. A good way to avoid this is by asking several why-questions to reach the problem's root cause. Here is an example. Suppose you want to enhance the on-time completion of the projects in your team. If you ask, "How can we finish the projects on time?" it will be too vague and broad to answer. So instead, you can first ask a few why-questions to find the real cause of the dealy. For instance:

- Projects aren't delivered on time. Why?

- Because they're not started on time. Why?

- Because the department tasked with the job is severely backlogged. Why?

- Because they're understaffed.

The questioning goes on until the problem is clear enough. This might sound simple but asking "why" can cause you to dig deeper in any situation. To answer a "why" question, you need to have a good idea about what's happening in that particular environment. We will review how to formulate original and inspiring problem statements in more detail later in this chapter.

4.   *Unclear constraints and expectations*

Trying to brainstorm successfully without knowing the constraints and expectations is like trying to win a football match without knowing where the goalposts are and how far they are from each other.

This fallacy goes hand in hand with believing in no structure and planning to stimulate creativity. Unfortunately, many people misunderstand out-of-the-box thinking for brainstorming. There are always boundaries of what is feasible and viable in terms of time, money, regulations, etc. This is the case for both individuals and companies.

The previous chapter discussed brainsteering as a variant of traditional brainstorming, which takes context, goals, and constraints into account. These constraints could include not going beyond a specific budget, limiting the available time, or targeting particular goals. These boundary conditions define the box of viable solutions. This might sound counter-intuitive to many. On the one hand, brainstorming requires creative and out-of-the-box thinking. On the other hand, successful brainstorming needs a carefully defined box of targets and constraints. How could we reconcile this?

The key concept here is to realize that brainstorming is more than the divergent phase commonly known by many people. Neglecting the convergent phase is so common that most people think of brainstorming as going to a meeting, sharing some mental explorations on the topic, and simply leaving, hoping that their ideas will miraculously solve the problem.

A successful brainstorming session consists of various divergent and convergent phases, each with its mindset and approach. Note that brainstorming also involves combining, refining, and clustering ideas. So while the raw idea itself may be disqualified by the limit you put, it could be refined to meet the requirements.

5. *Insufficient Orientation*

Have you ever been dragged away from your desk to attend a meeting and then asked to think out-of-the-box and find creative ideas? Without proper orientation on the problem, context, boundary conditions, brainstorming technique, and process, people cannot understand the issue and what to expect in the session. They tend to come up with uninformed thoughts, adopt a conformist view, and stay in good stand with the group.

Proper orientation sets the stage and tone of the sessions and guides the group towards reaching the desired goal. Orienting the group members should start even before the session begins by providing the essential information about the problem and the upcoming session. Once the session convenes, the facilitator should explain the issue further, introduce the brainstorming technique.

In addition, establishing the ground rules for brainstorming is an integral part of the whole process. Perhaps the most critical rule is the one that disallows criticisms in the divergent thinking phase. The last thing you want is for a single team member to start bursting out criticisms that somehow dull out the flow of the whole conversation. This is why the facilitator must impress the team members that criticisms should be left for later stages of the session when you've moved past the divergent thinking and started with convergent thinking. The facilitator has to be constantly on guard for any signs, peeps, or mumbles that would dim the results of the divergent phase. At this stage, the facilitator is really like a referee who makes sure that everyone in the playing field stays within the imaginary lines set by the game.

6.   *Failing to Have Everyone Contribute*

The whole point of group brainstorming is to use the mental power of the entire team and to achieve that, everyone should have the opportunity to contribute.

The issue could be more severe than people not talking. In other words, the brainstorming output not only suffers from more introverted members remaining silent but from reduced individual contributions as a whole.

Apart from individual personalities, the group dynamics and the psychological effects play a significant role in how successful the group brainstorming would be. Many studies show that individual brainstorming could generate more ideas than group brainstorming. We will discuss this counter-intuitive finding in more detail in the next chapter. The primary reason is that the group dynamics is not always constructive. On the contrary, even some of the knowledgeable and experienced members of the group could underperform, primarily due to the following two psychological effects:

- *Social matching*: This is the tendency of the participants to match the lowest level of productivity shown in the group. It is like regression to the mean, where a few cynical and uninspiring members bring the whole team down.

- *Social loafing*: This is when people work less hard in a group setting than they would do independently. People become more inclined toward social loafing if they feel their individual efforts do not matter for the group or are not valued enough.

The outcome of such psychological effects is that the more capable members of the team tone their contributions down. A common reason for this is when the brainstorming participants know that there are freeloaders around. For example, if the team lead or the session organizer takes the ideas and presents them to the higher management as their own.

Planning the session, proper orientation, and facilitation, using techniques offering more anonymity such as brainwriting will help improve the group dynamics. Other factors such as the company culture and the management are also essential, although more challenging to change.

7. *Settling for too few ideas too soon*

Brainstorming is an organized form of productive thinking, a deliberate effort to get as many ideas as possible, get them to have sex with each other, and hopefully create even more and better ideas. If we think of brainstorming as a Darwinian process of evolution and selection, then the number and diversity of the ideas are critical for a brainstorming session to thrive. The greater the number of ideas, the higher the chances that some will be fit to survive.

Prematurely stopping the divergent thinking stage of brainstorming is an excellent way to sabotage the session. Maybe you wonder how you could know if the divergent thinking should continue. If the participants quickly converge on an idea and try to justify that's the best possible idea or stick with the obvious ideas, the group should keep thinking.

In his book *Think Better*, Tim Hurson mentions how the first third of the ideas are usually the obvious ones in an ideation session. The second third starts to become more creative but still primarily mediocre. It is only in the third third of the ideas that our brains start to become creative since we've exhausted the obvious and semi-obvious ideas.

Sticking with the obvious solutions means that the group is still in the first third or maybe in the second third of the possible ideas. Of course, no one can guarantee that continuing the divergent phase will end in a breakthrough idea. However, just like a more diverse ecosystem has a higher chance of some surviving, a more comprehensive range of ideas enhances the odds of some becoming a breakthrough.

8. *Having an untrained facilitator*

By now, it should be pretty clear how critical it is to have a trained facilitator to run a group brainstorming session. Without a trained facilitator who knows everything we discuss in this book and understands the role of brainstorming in the creative problem-solving process, trying to have a successful brainstorming session is nothing more than a shot in the dark.

One of the biggest mistakes an organization can make when it comes to facilitation is to let the boss be the facilitator. In such as session, the participants will self-censor, and the boss will edit the ideas as they think is the right way. Furthermore, the managers are usually so occupied with the day-to-day running of the business that they can't properly prepare the meeting and follow up afterward.

9.   *Poor idea evaluation*

Most people tend to overlook the convergent thinking phase of brainstorming because they assume that brainstorming equals generating lots of ideas. Once the ideas are generated, the hardest part of the process is taken, right? Wrong! The outcome of such a mindset is that the participants end up with a load of ideas and go for the quick wins and ideas easy to implement.

The key here is once more understanding the fact that brainstorming needs at least one convergent thinking phase to sort, analyze, combine, and refine the ideas generated in the preceding divergent phase. The participants should receive a proper orientation that their work does not finish after having many ideas.

Ideas rarely appear in a perfect form. They usually need debate, analysis, and refinement. Withholding criticism is a fundamental brainstorming rule, but only in the creative thinking stage and not the entire brainstorming process.

Every brainstorming session should have at least one divergent and one convergent stage. In the convergent stage, the participants should critically evaluate all ideas against the boundary conditions defined at the beginning of the session. Within those boundaries, the ideas could be further refined and combined.

10. *No closure or follow-up through*

How does it feel to attend a brainstorming session, put the time and effort, and leave with no idea about the next steps? Or deciding on the future course of action only to see it being overruled by the real decision-makers? It would be frustrating and demotivating.

The participants should have a sense of closure at the end of the session. They should know what the next steps will be and how their contributions are used. One of the most important items is to explain if the group or management will select the final ideas. If the company management decides, the facilitator should clearly explain this to the group members and inform them about how their ideas will be handled.

The Ingredients of a Winning Brainstorming Session

If you are in charge of running a brainstorming session or planning to attend one, there are various strategies to help you get the creative ideas you are after. Some of these strategies are designed to avoid the common brainstorming follies, and others get the most out of the session. Put these tactics to work for you and get the odds of leaving the session with actionable ideas on your side.

Preparation

The work involved in a brainstorming session is not just the session itself. There are pre-session and post-session efforts that are essential to promote and execute successful brainstorming.

Most people think of brainstorming as only the session itself. Therefore, preparing for the session is often perceived as optional. But if you want to have a meaningful return on your invested time and effort, you need to establish the basic facts about your brainstorming session.

There are three groups of such facts:

- What to expect in the session
- Who to involve
- How to get the participants to think before the session

If you are the organizer of the session and not the facilitator, you need first to find a trained facilitator. If you will only attend the session, your job is mainly to digest the information on the discussion topic and think about the problem. For now, let's assume that you are the facilitator and want to set up a face-to-face brainstorming session. We'll discuss the ingredients of a successful virtual brainstorming session later in this chapter.

What to Expect in the Session

The most fundamental fact about a brainstorming session is the problem statement. This includes both recognizing the problem and phrasing it correctly, clearly, and engagingly.

Before rushing to the session, ask yourself what problem you like to solve and what question you will present to the brainstorming group. Do you remember the example of Granada Television and the advertising rights? Here is another example. Perhaps you know that ink is costly. Imagine you run a print shop or another business with high ink consumption, and would like to reduce the expenditure on ink. What would be the brainstorming question? An obvious one is "How can we print fewer pages?" But what if it is not possible to bring back the number of printed pages? In this case, the problem is high ink consumption. So, the real question is, "How can we save ink?" A Dutch company called SPRANQ came up with an innovative solution. They developed Ecofont; a unique TrueType font with tiny holes in the characters resulting in significantly fewer printed pixels and up to 50% saving in toner ink. The UK stationary retailer Ryman followed the same idea and developed another ink-saving font called Ryman Eco containing thin lines with hollow spaces in between. Rayman Eco consumes one-third less ink than standard fonts such as Arial or Times New Roman. To give you a feeling of how much of a difference that can make, if everyone uses fonts such as

Ryman Eco, it can save over 490 million ink cartridges, cutting over 6.5 million tons of CO_2 every year.

Once you figure out the exact problem, the next step is phrasing the problem engagingly. A good technique is writing variations of the same question over and over again. You want to present the question uniquely so that it also inspires unique answers. If you're unsure, here's a checklist of what a proper question should sound like:

- *It addresses just one issue.* If several problems are on the agenda, make sure you separate them into individual questions. This is important because you want to be able to address each issue individually. Doing so helps you approach the problem more clearly, especially when considering different factors. You don't want the factors of problem A to somehow mess up with how problem B is answered or vice versa.

- *It is specific.* A broad question will leave the participants confused and drowning in data. However, the question should not be too specific that limits the creative thinking of the team.

- *It should present the issue in a constructive and solvable way.* For example, instead of saying, "How can we make sure that we never have late deliveries?" you can say, "How can we make sure that all our deliveries are on time?" This helps adopt a positive mindset dedicated to achieving what you want versus avoiding something you don't want.

- *It should be actionable.* For example, the "how might we…" questions put the onus on us. The question is directed to the group and asks for actionable solutions from them. The goal is to eventually find answers that you can implement because it is within your control.

Let's face it, we constantly watch what we say in the meetings not to look dumb or cross the lines. Hence, it is crucial to open the brainstorming session with a question that helps everyone feel safe to share their thoughts. In fact, the atmosphere of the entire session should reinforce that feeling of safety. Otherwise, the participants could shut down.

Asking engaging questions is a complex skill to master and a vital one for productive brainstorming. The problem statement should not be too narrow and therefore limiting the participants' creative thinking. But, on the other hand, it should not be so broad that it becomes difficult to answer.

There are many ways to formulate problem statements. You can start the session with more straightforward and more descriptive questions to more complex and analytical ones. For example, start with open-ended observational questions to set the stage for the actual problem statement. Observational questions could be answered based on feelings and experiences and are great to kick-start the session and engage everyone. Here are some examples:

- What are your first thoughts when you look at this product?

- What do you think of our website?

- Which app interface do you like more?

Once there is some engagement, you can pose the main problem statement. You can use the principles mentioned above to develop great brainstorming questions in every discipline. Here are some examples:

Business Growth

- How can we double our yearly profits?
- How can we attract 10,000 new members?
- How can we get 50 more clients?

Developing New Products and Services

- How would we expand our services to new markets?
- How can we position our product to take advantage of the market demand?

Cost Saving

- How can we cut the production time of product XYZ?
- How can we cut the cost of product XYZ?
- How can we lower our overhead costs?

Social Responsibility

- How can we cut the carbon footprint of our production?
- How would we create more eco-friendly packaging for our products?
- How would we encourage our clients to opt for eco-friendly packaging?

Process Optimization

- How would we speed up delivery?
- How would we improve our storage capacity?
- How can we strategize for a work-at-home situation?

Employee Satisfaction

- How could we reduce department workload?
- How could we encourage employees to show up at work on time?
- How could we improve employee satisfaction?

Communication

- How can we streamline communication so that it's faster and more transparent?
- How can we quickly inform employees about any changes in the company regulations?
- How can we encourage employees to use new software tools to increase communication efficiency?

Recruitment

- How can we attract young and talented employees to our team?
- How can we retain talented employees?

- How do we build a strong team for our projects?

If you do not see enough engagement with your initial problem statement, do not give up! Instead, throw a wrench into the session by asking some lateral thinking questions. These might sound surprising and even sometimes jarring to most people. However, the intention is to force participants to think about the problem differently. Here are some examples:

- What if we had infinite resources to do this project?

- What if we adopted an entirely new approach?

- What would happen if we didn't do this project?

Asking questions is an integral part of every brainstorming session. As the brainstorming winds down, the facilitator should try to guide the group with some actionable questions to bring some structure to the otherwise messy and frantic brainstorming session and move. For example, questions such as "Which of these ideas should we execute first?" or "In your view, which ideas are the most viable?"

Choosing the right brainstorming technique is another important factor. You can go for a virtual session or one of the face-to-face methods discussed in Chapter 4 based on your situation, the number of participants, the sensitivity of the topic, available time, etc. You can also combine various techniques to have a more effective ideation process. For example, encouraging the participants to think up ideas on their own before coming to the group session is a great way to have both individual and group brainstorming benefits. We'll talk more about this later in this chapter. Having an initial virtual session before getting together, discussing the topic face-to-face, and dividing the brainstorming process into an ideation session and an analysis or discussion session are some of the other options.

The organizer or the facilitator should also think about the meeting duration, how to handle the outcomes, and who decides on the final ideas (I.e., the brainstorming group or the company management), whether to have a follow-up session, among others. How detailed these preparations should be, depends on the brainstorming topic and its context. In any case, you need to have at least a general idea of the process you would like to follow and take preparatory steps if required.

Who to Involve

Participants should be invited based on the contribution they can make. In addition, they should be willing to participate and have a genuine interest in the topic. It will also help if they are directly involved with the problem. The group should ideally have a wide range of backgrounds and a good mix of experts and non-experts. The number of people in the group should be large enough to have the breadth of thoughts but not so large that the session becomes chaotic, and people can get by without participating. In most cases, 5-10 people is an ideal number.

At least some of the participants should have first-hand knowledge and experience on the topic under discussion. Furthermore, having certain personalities in the team is desirable. An example of such personalities is the devil's advocate. The devil's advocate is the individual who is constantly voicing doubts. If everyone is saying YES, the devil's advocate would be the one saying no. On the other hand, if everyone is saying NO, the devil's advocate would be the one saying WHY NOT? You might wonder if this could bring the ideation process to a halt. Not exactly!

Have you ever noticed how the submission of a pretty good idea somehow changes the mindset of everyone in a team? Suddenly, this idea becomes the only idea; everyone reaches a consensus and stops looking for a new idea. This effectively eliminates the creativity process and locks your participants into a solution. This is where the devil's advocate becomes incredibly useful.

With someone who expresses doubt and doesn't always accept the majority decision, the devil's advocate can keep the conversation open. By disputing the logic of the group, the devil's advocate encourages everyone to think the idea through and assess it from different perspectives. While this might seem like a bad idea because it prolongs the discussion, having one in your team can help ensure that the outcome goes beyond the obvious.

The brainstorming group should preferably include diverse backgrounds. In a homogenous group, everyone thinks similarly and reaches similar conclusions, or they just agree with each other. A homogenous brainstorming group tends to diverge to a few final ideas quickly. Diverse backgrounds enhance the chances of the diversity of thoughts, which could profoundly increase group productivity.

How to Get the Participants to Think Before the Session

Inviting a group of people to a meeting in general and a brainstorming session in particular without providing sufficient details and context about the meeting will make them feel confused and unwelcome. By contrast, attaching an agenda to the meeting invite, including the problem statement and context, and some ground rules will allow the participants to prepare for the meeting.

You can also provide the constraints and the opportunities the team faces, any relevant history, why solving the intended problem matters, and why you choose the specific brainstorming technique.

Your goal is to get the participants' buy-in prior to the meeting so that everyone comes prepared and ready to engage. You encourage the participants to start thinking about the problem even before the session begins by providing sufficient details in advance. This is like having an individual brainstorming by each participant before joining the group session.

A good facilitator helps participants to be at or close to their best during the session. Most people need time to research and understand the problem. That is why informing and engaging the participants in advance are so important. In practice, most successful brainstorming sessions are occasions to share and refine partially worked-out ideas. To find such ideas, participants need information and time before attending the group session.

Brainstorming is an organized form of group creative thinking. If creativity is domain-specific (as discussed in Chapter 1), successful brainstorming requires specific knowledge and expertise to solve the problem. The notion of simply putting a group of people together and expect them to generate breakthrough ideas is very simplistic, especially for complex business problems. If you want to brainstorm some themes for your kid's upcoming birthday party, you probably just need to sit down with your partner or some friends, chat about a few options, and pick one. However, suppose you are looking for ideas to develop a winning marketing campaign, a breakthrough consumer product, or an investment strategy. In that case, the team will need a lot of market research, one-to-one convos, focus groups, prototypes, etc., before they can develop impactful ideas.

Too many organizations and their managers fail to understand the facts that productive brainstorming needs preparation, and brainstorming, in general, is just one part of the problem-solving process. Sometimes group brainstorming becomes a management crutch, i.e., an easy and painless way for the management to pretend that their people are involved in the problem-solving process. Poor preparation is widespread in such brainstorming sessions.

Running the Session

Finally, it's time to get together and have a group brainstorming session. With a well-prepared session, everyone has a good idea of the discussion topic and its context, the goals and constraints, the brainstorming method, the team, and some basic ground rules. The only remaining step is that the organizer or facilitator of the meeting should arrange a suitable room and check if the basic tools are available.

It is not practical, realistic, or even helpful to propose a standard procedure to run brainstorming sessions. Every session is unique and might need a slightly different approach. In this section, we'll review some general guidelines to have productive group brainstorming sessions. The facilitator and participants should be ready and willing to adapt the course of session on the fly.

Opening and Navigating the Session

Before generating and discussing ideas, it is better to spend some time reviewing what was shared in preparation for the session. The facilitator can provide further details and answer any questions on the problem statement, brainstorming technique, or the ground rules. It is similarly important to create a positive and relaxed mindset to help people be at their best. For instance, solving a couple of lateral thinking puzzles is a good way to encourage creative thinking in a playful manner. Here is an example: Three people enter a room but only two walk out. The room is empty. Where is the third person? Doing such exercises is a safe way to help the participants distance themselves from the usual linear and analytical thinking. By the way, the answer is that the third person was in a wheelchair and wheeled out! Or maybe the third person was hanging down on a rope, like Tom Cruise into the vault scene in Mission Impossible! These ice-breaker exercises are among the best options to relax the group and kick-start the creative thinking process.

The brainstorming process has two major phases. Divergent thinking dominates the first phase, and convergent thinking rules the second one. Every phase needs its specific mindset and approach. Withholding criticism is the main principle in the divergent phase. We discussed the other principles in the previous chapter, such as recording all ideas, eliminating the obvious ideas using braindumping, and combining and refining ideas.

Once the stage is set, it is time to dive deep into the divergent phase and proceed with the session as planned and discussed with the team. You never know how the session progresses since every session is unique. However, a well-prepared session has high chances of engaging and productive discussions. Nevertheless, the facilitator should be ready to boost the discussions if needed. The facilitator should also manage the time, keep an eye on the group dynamic and interaction, and properly warp the session up and follow up with the team. Let's discuss each point in more detail.

Time management

What is the optimum runtime for a brainstorming session? The short answer is, "It depends!" The long answer is that the facilitator should set the time depending on the brainstorming group, the discussion topic, and circumstances. For example, a short session (e.g., 30 minutes) might work for a deadline-driven team, but not for everyone. If the topic is complicated or the group individuals are not accustomed to thinking creatively, imposing a short deadline leads to braindump ideas. These are the kind of ideas that are logical, obvious, and predictable. Given a short time for creative thinking, most people will instantly think about the obvious so that by the time they're done eliminating the obvious answers, there's no time left for the creative mind to take over.

On the other hand, a long session could leave people lax and feeling as though they have all the time in the world. So how do you find a happy medium where the time limit bolsters creativity instead of killing it? A good approach is allotting time for specific stages, such as 15 minutes to frame the problem, 15 minutes to think of the obvious ideas, 30 minutes to think of creative ideas, and 30 minutes to evaluate them and wrap up. This usually works better than running a long session, covering all stages continuously.

The facilitator should explain the time slots, therefore allowing each member to create a map in their head about ideas and the time frame they're required to come up with them. Longer meetings will need periods of rest in between the stages.

Some facilitators also try to pair timed sessions with a quota. The goal is to have the participants develop a specific number of ideas within the allotted time frame. By doing this, participants are forced to squeeze out as many ideas as possible to meet that requirement. The beauty of this method is that since brainstorming encourages the crazy and out-of-the-box narrative, many participants would actually choose to suggest something insane to meet the required number.

If the topic is complex, the brainstorming group is large, or the discussions are going on so well that it is worth extending the time limits, the facilitator could restructure the session or plan a follow-up session. This is not ideal but sometimes is unavoidable.

People Management

Time management is not the only skill to keep the meeting on track. Another important skill is managing group dynamics and interpersonal interactions. To get the most out of the session, the facilitator should consider the different personalities in the session and how to promote good interactions and high-quality participation. Let's first review some general personality traits and how to deal with them in a brainstorming session.

Extroverts usually like group brainstorming; it gives them energy. However, for many introverts, attending a traditional group brainstorming session is a dreadful experience. A group of people continuously talking and even fighting for the airtime is a painful experience for them. The introverts could be as creative as the extroverts, they just need a different setting.

Brainstorming techniques such as brainwriting, brainwalking, or the stepladder offer a certain amount of anonymity and tranquility that appeals to introverts. Similarly, sharing the problem statement and context in advance allows the introverts to think up their ideas on their own and without being overwhelmed by the group discussions.

You can even consider virtual brainstorming as a stand-alone process or in preparation for a face-to-face meeting. Again, introverts might find comfort in the virtual process.

Some people like to mull over the problem and develop their ideas meticulously. Furthermore, regardless of being an introvert or extrovert, not everyone is a fast thinker who feels comfortable thinking up ideas on the spot. Good preparation for the meeting and allowing at least a few days to understand the question and context could greatly help the whole team engage better during the session.

Choosing a suitable technique, dividing big groups into smaller subgroups, and allowing the participants to think before coming to the meeting will create a solid structure in the divergent phase of the brainstorming process. Similarly, such measures will prevent any individual from dominating the session or being disengaged. However, once the convergent phase begins and the team starts evaluating, combining, and refining ideas, the group interaction could go out of hand. Therefore, in this phase, the facilitator should be more strict with directing the team, calling out the time, and sticking to the plan.

Wrap-up and Follow-up

Closing the session is just as important as opening it When completing a session, you want to leave the participants with an impression of what was achieved and what happens next.

Evaluating ideas is a critical step to have a proper closure. This is the convergent phase of the brainstorming process and is sometimes overlooked. Knowing the targets and constraints is essential for evaluating the ideas. It would help if the facilitator shares the targets and limitations in advance and reiterates them at the beginning of the session. For example, if "Time to Market" and "Profitability" are among the evaluation criteria, the team can discuss whether each idea will be profitable, to what extent, and within what time frame.

A practical and straightforward approach to rank the ideas is to divide them into three categories: ideas that meet the requirements, ideas that don't meet all the requirements but are still interesting, and everything else.

Once all ideas are discussed and divided into these three categories, the facilitator should explain if the team or the leadership will choose the final ideas. If the leadership makes the final decision, the results should be communicated to the brainstorming team.

Running an Engaging Virtual Brainstorming

With the recent trend of remote working, virtual brainstorming has become a valuable tool for many organizations. In some cases, it could be the only real possibility to engage team members in different time zones and geographies.

Virtual brainstorming is more favorable for introverts, and it is easier to avoid some of the typical brainstorming pitfalls such as dominant personalities or production blocking. Furthermore, the asynchronous nature of virtual brainstorming allows the participants to do their individual brainstorming and then join the group session. These benefits could be blessings for the brainstorming process, and you can use them to your advantage.

Running an engaging virtual session follows almost all of the same principles as for a face-to-face session. Preparation, time and people management, and proper closure and follow-up are all essential.

The most significant difference is introducing the virtual brainstorming tool and process. The facilitator should explain the tool and how to use it, the brainstorming process, and the ground rules. For example, the process could consist of an introductory video conference, an asynchronous period of individual brainstorming using the virtual tool, and then another video conference to discuss the ideas.

A unique characteristic of virtual brainstorming is that the outcome could be abstract and not actionable. A significant body of research data shows that when people are socially or in time or space distant from a topic, they tend to think more abstractly about that topic. This is called *Construal-level Theory*. Since participants are physically remote from the site of the problem, they can think more abstractly. Although abstract thinking could be beneficial to find insights and analogies and to remember past experiences, it can prevent people from generating specific and actionable solutions.

The facilitator should remind the participants about this issue, and everyone should watch out for endless abstract discussions. Virtual brainstorming should contain divergent and convergent phases, just like face-to-face brainstorming.

Hence, virtual brainstormers should evaluate ideas critically and try to find detailed and specific solutions.

Virtual brainstorming eliminates production blocking, enables the feeling of anonymity, and enhances the delivery of new ideas by reducing the exposure to other participants' ideas in the divergent phase. A well-curated virtual brainstorming session retains the power of collective thinking and can go beyond traditional brainstorming.

In 1988, Peter Drucker (one of the greatest management consultants of the past century) predicted that technology would transform teamwork and collaboration. His prediction is coming true in many fields of business, technology, and education. For example, technology is making brainstorming more effective by replacing verbal and physical sessions with written and virtual ones. It would be unwise not to join this transformation.

One thing that will not change about brainstorming in the foreseeable future is individuals thinking up ideas. Individual brainstorming deserves a separate discussion, and that's what we will do in the next chapter.

Chapter 6: Individual Brainstorming

"Whosoever is delighted in solitude is either a wild beast or a god. He never is alone that is accompanied with noble thoughts."

-Aristotle

Solitude is valuable for creativity. For many people, especially introverts, solitude is the time to focus, reflect, think deep and access their memories and experiences. Far from constant stimuli, the mind could become self-aware, and creativity could flourish. Many artistic masterpieces and scientific breakthroughs were made after long hours of solitary thinking and working. So, how about solo brainstorming? Could it be superior to group brainstorming? Isn't that against the whole premise of brainstorming?

Alex Osborn famously claimed that his brainstorming groups produced 50% more creative ideas than if the same group members worked independently. Yet, after seven decades of group brainstorming, many people experience brainstorming sessions not remotely as positively as Alex Osborn did.

In recent decades, a growing body of scientific studies contradicted Osborn's claim. These studies show that group brainstorming could be significantly less productive than individual members thinking on their own. For example, a meta-analytical review of 800 teams by Dr. Brian Mullen and his co-workers at the Department of Psychology of Syracuse University in the early 1990s indicated that individuals are more productive idea generators when left to think independently. Furthermore, they found that productivity loss was the biggest for larger brainstorming groups, closely supervised teams, and when ideas are shared verbally and not in writing.

Such findings come as no surprise since we know that traditional brainstorming could be plagued by social loafing and matching, regression to the mean, production blocking, evaluation apprehension, and several other pitfalls. By now, you also know that new variants of group brainstorming, such as brainwriting or brainsteering, could effectively address most of these shortcomings. Nevertheless, the case for individual brainstorming is still strong and especially appealing to introverts and lots of creative people.

In Chapters 1 and 3, we discussed the thinking processes of the most creative humans of all time, such as Picasso, Einstein, and Darwin. None of these people attended a group brainstorming session to come up with their ground-breaking ideas!

But is that a good argument to say goodbye to group brainstorming and hole up alone at home or office? In this chapter, we'll find out if creative thinking is like solo dancing, Ballet, or both!

Why not Just Group Brainstorming?

Brainstorming started as a group activity and has remained so to a large extent. Collaborative thinking feels intuitively right; that a group of people could achieve more than their individual members. At least, that is the theory. In practice, group brainstorming could work well, but it requires a trained facilitator, careful selection of the participants, and meticulous coordination of their efforts. Unfortunately, not everyone is as dedicated as Alex Osborn to organizing and running brainstorming sessions. Hence, the output could be less than favorable.

You might wonder, given the pitfalls of group brainstorming, why is it still practiced so widely? There are three main reasons:

1. The new brainstorming techniques (such as brainwriting) are more effective than traditional brainstorming. They address most of the drawbacks but still require skillful facilitation and a lot of coordination.

2. With the increased specialization of labor, organizations need to bring people with different backgrounds to work

together. Remember, creativity is domain-specific and innovative solutions need collaborative thinking of people with varying types of expertise.

3.   Even if groups do not generate more or better ideas, engaging the stakeholders in the thinking process improves ownership, and the enhanced buy-in makes the implementation easier

Group brainstorming will remain an essential ideation technique, and it might be further improved using new digital tools and innovative approaches. However, some people would like to be left alone for good reasons, and the drawbacks of group brainstorming will not entirely disappear.

A fundamental misconception here is considering individual and group brainstorming as two competing techniques. Creativity psychologists know that creativity needs both solitude and collaboration. So why not adopt the same mindset for brainstorming? Solo thinking could be an asset on its own or to boost group brainstorming.

There are valid reasons to pay more attention to individual brainstorming as a creative ideation approach. Some people, specifically introverts, perform better when given a chance to generate ideas alone and in the comfort of their own company. But introverts are not the only ones that can benefit from solo thinking. The human mind functions better when given some solitude and the chance to wander. Ever wondered why some of your good ideas come in the shower? It's not just because of the incubation period but also due to the benefit of solitude. You can study concepts in your head, turn them upside down, and make conclusions and connections that are more difficult when distractions surround you.

Individual brainstorming doesn't require the input of another person. You go through the whole process on your own, and there is little or no evaluation apprehension. Who else will tell you that your idea is stupid if you're all alone? This leaves you more open to weird, crazy, and unconventional thoughts, giving you the freedom to entertain wild concepts that you might be afraid to share when working as part of a group.

Many professional environments tend to over-emphasize group work. Similarly, they under-rate individual thinking and the role of introverts in their organizations. Recognizing the value of solo thinking and incorporating it into the extrovert-dominated group brainstorming culture could enhance employee satisfaction and productivity of ideation sessions.

Individual brainstorming could help us become better problem-solvers in our personal lives. It could also improve the quality of the group brainstorming session. The ideal scenario is to combine individual and group brainstorming in various phases of the idea generation process. This chapter will discuss how to brainstorm on your own effectively and how to combine individual and group brainstorming.

Individual Brainstorming Techniques

Individual brainstorming follows the same general principles as group brainstorming and consists of divergent and convergent thinking stages. However, there are also some significant differences. For example:

- It does not require extensive preparation as in group brainstorming.

- You are simultaneously the organizer, facilitator, and participant.

- You define the brainstorming question, choose the technique, set the time limit, decide when and where to brainstorm, etc. Hence, individual brainstorming is much more flexible and convenient than group brainstorming.

- The typical pitfalls of group sessions are absent. So, you don't need to worry about social anxiety associated with presenting ideas to a group of people or others rolling eyes at your ideas.

- It has a much looser structure. You can think up some ideas, leave the topic for a while, go back and generate more ideas, wait for a few days, evaluate the ideas over a period, etc. Then, you can draw a sketch or make a prototype of the idea and get advice, repeat the divergent-convergent cycle all over again, etc. The flexibility and freedom are some of the most significant advantages of individual brainstorming and the main reason that solo thinking appeals so much to creative artists, poets, writers, and polyglots.

Individual brainstorming consists of the following general steps:

- Formulating and presenting the problem to yourself.

- Generating ideas and recording them legibly on paper or digitally without editing.

- Once you're done with the divergent thinking phase and have recorded as many ideas as possible, leave them for a while, maybe for some hours or even days. Meanwhile, the ideas will incubate and mature in your mind. If you get a flash of inspiration in the meantime, make sure to record it too.

- Come back when you're feeling refreshed and incorporate any thoughts you might have had into the notebook.

- Combine, cluster, and refine and evaluate the idea.

- Select the final thoughts or repeat some or all of the previous steps.

There are four main techniques to execute these steps in a structured way. Of course, you can combine or tweak these techniques to best fit your needs and circumstances. After all, you are the organizer and facilitator of your thinking session!

Free Writing

Here you first write down any idea that comes to your mind on the brainstorming topic. Do not care about spelling, grammar, or writing complete sentences. Instead, focus on putting as much information as possible on the paper (or in a digital file). The ideas could form continuous prose, with one idea or piece of information leading to the next one. Once the divergent thinking phase is complete, you have a disorganized piece of writing which needs analysis and screening, i.e., convergent thinking.

If you like to try free writing, you can follow these steps:

- Write the brainstorming topic on a piece of paper on in a digital file.

- Write down anything coming to your mind on the topic. Try to include as many ideas, supporting facts, details, and examples as possible. You can write them down as you wish; in bullet points, diagrams, graphics, single words, whole sentences, etc.

- Re-read your text and highlight or circle anything that seems interesting.

- Organize, combine, and expand the selected passages until reaching some concrete ideas.

- Repeat some or all of the previous steps if needed.

Free Speaking

This technique closely resembles free writing. But instead of writing the ideas, people record their own voices while thinking out loud. Some people find this easier and faster.

First, give yourself a set time to think and as many ideas as possible, speak them up, and record them. Then, listen to the recording and write down any valuable clues. Finally, refine and combine the clues until having satisfactory results.

Mind Mapping

We discussed mind mapping as a virtual and face-to-face group brainstorming tool. You can also use it as a visual note-taking technique to diagram your thoughts and their connections. It creates a spider web of ideas with the brainstorming question in the center.

A mind map keeps the ideas organized. It grows from the core question like branches of a tree, and each branch contains bubbles with ideas inside and sub-branches with relevant details. You can first generate the ideas and organize them in the mind map. Then, review them, add details, connect and combine them and revise and expand the map until satisfied.

Word Association

In his book, *Thinkertoys*, Michael Michalko compares the thinking patterns of the human mind to a tetherball that can't escape from the pole. We tend to follow our habitual thinking patterns unless external stimuli kick us into a new way of thinking. For example, associating unrelated words is a simple yet powerful tool to force our minds to make new connections and find potentially creative ideas.

There are two primary forms of brainstorming using word association:

1. *Structured word association:* This is a loose form of mind mapping. It starts with writing a summarized version of the

brainstorming question on the paper. Then write down a few terms related to the main question and continue jotting down the first words coming to mind when you hear each of the related terms. The outcome will be several word chains. Next, reread the word chains, write down your reactions, and look for recurring themes or ideas worth exploring.

2. *Random word association:* The goal is to force connections between the brainstorming topic and unrelated random words to provoke new connections. To execute this method, first, draw a table with two columns. Next, choose the root question word and write a list of its associated words and phrases in one of the columns. Then select another root word entirely unrelated to the first root word, and add its associated terms and phrases to the second column. Finally, pair up phrases from the two columns and see what concepts or clues come from these combinations.

The "Cut-up" Technique

Sometimes combining various ideation techniques works better. A good example of such hybrid methods is the "Cut-up" technique, which has roots in the Dada art movement of the 1920s. It was popularized in the 1950s and 60s by artist Brion Gysin for creative writing and was later adopted and refined by American novelist William S. Burroughs and English singer-songwriter and actor David Bowie.

The Cut-up technique is a blend of free speaking and word association. For example, Bowie cut out random words and phrases from newspapers and his hand-written notes and mixed and rearranged them to form associations. This helped him to spark new ideas and themes for his music.

If accomplished people like David Bowie and William Burroughs could benefit from the Cut-up technique, maybe you can try it out too! In brainstorming, you can use this technique to make structured and random associations to generate new ideas. Here are the steps:

Step 1: Record and Transcribe Your Thoughts

You have framed the right question and are ready to start the ideation stage. Imagine you have a one-on-one conversation with yourself! Express your thoughts on the topic out loud and record your speech. Talking out loud to yourself has tremendous power. We usually share more thoughts when speaking than when we write. The recording process also guarantees that no quip or fleeting idea is ignored.

Spontaneity is essential when it comes to brainstorming. Hence, you might find yourself floundering a bit, backtracking, mumbling, whispering words to yourself, or perhaps even going off on a tangent while trying to explain your ideas. These are all perfectly normal. Once you're done with talking, transcribe your speech, and add notes to unscramble your thoughts.

Step 2: Cut Them Up

Yes, you're supposed to cut up your transcribed papers into chunks and pieces. You can make random cuts or cut out phrases and text fragments. Don't feed the transcript through a shredder because this will stop you from enjoying the next step!

Step 3: Piece them Back Together

You can put all the pieces in a bowl or lay them out on a table to see the whole picture. Then try to rearrange the text fragment and make new connections and combinations.

If you prefer free writing or visual associations more than speaking and recording yourself, you can replace the free speaking step with free writing or even with a digital brainstorming. To bring the Cut-up technique to the digital age, you can use virtual boards such as MURAL to collect bits and pieces of ideas and rearrange them to random or structured combinations. In addition, Digital tools allow you to drag and drop shapes and notes and make the brainstorming process more visual. They also make it possible to engage in a group "Cut-up" technique.

Salvador Dali's Brainstorming Method

The renowned Spanish Surrealist painter Salvador Dali followed one of the most unorthodox techniques to harness his creativity. While there's no guarantee that this would also work for your individual brainstorming, there's no harm in trying.

Dali is best known for his unreal or unpredictable juxtaposition of objects to create surreal imagery, such as his 1931 painting *The Persistence of Memory*. One of his favorite techniques to get ideas and inspiration was to capture images that appeared to him in the state between sleep and wakefulness. He experimented with different ways of capturing those surreal imageries. One of them was sitting in a chair next to a tin box, holding a spoon over the box. He then tried to relax his body and start falling asleep. When he began to doze, the spoon would drop inside the box, waking him up to capture what he saw in his short dream.

You can adopt Dali's method for brainstorming using the following steps:

- Get yourself a chair, a box, and a spoon, just like Dali did.

- Think about the problem at hand, its context, your ideas, constraints, etc., while sitting on the chair, holding the spoon above the box.

- Stop thinking, relax your body, and calm down for a while. When you start to doze, drop the spoon and wake up to capture some images.

- Record whatever you saw before waking up.

- Look for associations and connections.

Start asking yourself to narrow down the ideas for brainstorming purposes. Some questions you can ask include:

- What did I see?

- How does this relate to my problem?

- Is there any relationship with my challenge?

- What is out of place in that picture?

- What do I don't like about what I saw?

- What do those images remind me?

- How are these images representative of the solution to my problem?

Imagine that a restauranteur uses Dali's method to brainstorming new promotion ideas. The restauranteur wakes up and keeps seeing giant food images. The associative link could be to use the food itself as a promotion. So, he decides to offer various free food items every day. He advertises the free food with neon signs. But the customers don't know which item is free until they inside the restaurant and order something.

Dali's technique could offer a unique divergent thinking tool to kick start the creative ideation process. The outcomes could become inputs for other methods such as free writing or word association. You never know how effective these techniques are until you try them!

Stimulating Creative Thinking

Individual brainstorming is when you engage your creativity to generate ideas. in a relaxed, informal, and targeted way. Brainstorming techniques, such as freewriting or mind mapping, are great for guiding and organizing your thinking process and recording the ideas. But eventually, it is creative thinking that makes the difference between getting brilliant or mediocre ideas. Moreover, when you are trying to develop ideas on your own, creative thinking is even more essential since you cannot rely on the synergy between the group members to build on each others' ideas.

It would be simplistic to assume that we can follow conscious and deliberate processes to stimulate creativity because creative thinking does not happen entirely in our conscious minds. Forcing our rational and analytical minds to think hard about the problem is the easy part. But it is insufficient and even counterproductive. To truly stimulate our creativity, we need to try various approaches to allow our conscious and unconscious minds to make associations and get inspiration from whatever sources they can.

We started the book by dissecting creativity. Now, it's time to put creativity in a more pragmatic context. We will review five approaches to leverage the power of creative thinking to find practical ideas within the boundaries and limitations of real-world situations. Of course, nobody can predict or guarantee the outcome of a creative thinking process. What you can control is to stack the odds of finding breakthrough ideas in your favor.

Organize Your Thoughts

Let's start with the most straightforward approach. You might tackle a problem or a question on multiple occasions and over some time. Ideas could come to you at any time during this period. Hence, it is essential to record and organize them properly. You can use one or more of the individual brainstorming techniques discussed earlier in this chapter to keep track of the ideas, spot inconsistencies in your line of thinking, combine and refine ideas, etc. For example, mind maps are great tools to generate ideas using associations, and a freewriting journal is a simple and effective tool to write and review ideas at any time.

Ideas inspirations are fleeting. So, write them down as soon as they strike your mind. You can use the journal to practice thinking freely. Who knows, maybe some of those ideas will become viable leads! This approach is more about consistency, discipline, and establishing a solid ground to build your ideas.

Reframe the Problem

A common trait of creative people is that they reframe the problem more often than less creative people do. What does reframing mean? According to Thomas Wedell-Wedellsborg, the author of *What's Your Problem*, reframing is a crucial and underutilized creative problem-solving skill. It is not analyzing the problem or even finding the real problem. Instead, it is solving the right problem, i.e., shifting the way you see the problem. Reframing is so vital because it determines the scope of the potential solutions.

Wedellsborg gives several examples of reframing in his book. A classic one is the "slow elevator" example. Imagine you are the owner of a building with an old and slow elevator. Your tenants constantly complain and even threaten to cancel their lease contracts. In your eagerness to solve the problem, you frame it as "How to make the elevator move faster?". Hence, you focus on solutions such as upgrading the motor or even replacing the elevator with a more modern one.

Reframing is taking a pause and re-thinking the problem. In the "slow elevator" example, you can reframe the question from "How to make the elevator move faster?" to "How to make the wait time more pleasant and engaging" The range of potential answers immediately changes. Instead of costly upgrades on the elevator, you can refurbish the elevator interior, add some mirrors or play music!

Reframing also helps adopt a more positive and resilient mindset, encouraging us to keep working when feeling stuck with a problem or facing life adversities. For example, it will make a big difference if you consider an entrepreneur trying hard to establish their business while struggling with keeping a healthy work-life balance as a "bad father" or as "coping well in challenging circumstances."

Reframing does not change the essential nature of the problem. Instead, it redefines our perception of them. One of the most common applications of reframing is being creative while thinking *inside* the box. Idea generation in the real world always happens within certain constraints. Whether you consider those constraints as obstacles or opportunities makes a massive difference in how you approach a problem.

Countless stories of creativity occurred by reframing obstacles to possibilities. Ernie Schenck, the author of *The Houdini Solution,* gives several examples of creative problem solving by reframing in his book. One of these examples is the advertising campaign of the Mexican restaurant chain Chevys in the early 1990s in America. Chevys prided itself on its "Fresh Mex" food, and they wanted to communicate that message in their ad campaign. So they turned to San Fransisco agency

Goodbye Berlin & Silverstein to produce an ad that emphasizes their daily fresh menu. The issue what that Chevys had a very tight advertising budget

The creative team at Goodbye Berlin & Silverstein came up with the idea of going into the streets and asking ordinary people if they knew how fresh Chevys food was, got some funny answers, edited the recording, and aired the ad that same night. The message was if even the TV commercial is so fresh, imagine what the food would taste like! The ad campaign worked like gangbusters. The creative team at Goodbye Berlin & Silverstein used the budget limitation in their favor, though creatively inside that box, and came up with a superb solution.

Reframing usually involves challenging assumptions. In the example of Chevys, most advertising professionals will start thinking of well-known commercials from McDonald's or Coca Cola, assuming that they need to shoot for several weeks with a renowned director and some superstar celebrities on a tropical island. However, a whole new world of opportunities opens once you pause, question the assumptions, and redefine the question.

For instance, imagine you want to divide a pie into eight pieces with only three cuts. This riddle puts you inside a box and still asks for a creative solution. Most people drive themselves crazy by making three straight cuts in all kinds of ways. None of them results in eight cuts. But who said you could only use straight cuts?! Nobody, you assumed that. If you ask yourself, "What if I can make a round cut?", then you can quickly realize multiple ways to divide the pie into eight pieces with only three lines. For example, you can divide the pie into a ring and a circle with a round cut and then have eight pieces with two straight cuts.

Self-imposing restrictions could even be an effective way on its own to stimulate creativity. Take the example of the American children's author Dr. Theodor Seuss when he wrote *Green Eggs and Ham*. After completing *The Cat in the Hat* with 236 words, his publisher challenged Dr. Suess that he cannot ever break that record and produce an even shorter book. Dr. Suess won the bet by writing *Green Eggs and Ham*, one of the best-selling children's books of all time, with only 50 words!

Experience Creative Flow

You know that feeling when you are so absorbed in your work that you don't feel the passage of time? When you enjoy creating and feel both calm and excited? Mihaly Csikszentmihalyi (pronounced Me-High Chick-Sent-Me-High), one of the most prominent psychologists of our time and the father of positive psychology, calls that state of mind "creative flow." He explains such moments in his popular 1990 book *Flow: The Psychology of Optimal Experiences* as "

"A state in which people are so involved in an activity that nothing else seems to matter; the experience is so enjoyable that people will continue to do it even at great cost, for the sheer sake of doing it."

Csikszentmihalyi concludes that creative flow is highly correlated with exceptional creative performance and encourages us to try to experience the focus and inspiration to in the moments of flow. He says:

"The best moments in our lives are not the passive, receptive, relaxing times. The best moments usually occur if a person's body or mind is stretched to its limits in a voluntary effort to accomplish something difficult and worthwhile."

How can we expose ourselves to the moments of creative flow? Csikszentmihalyi provides several guidelines. Here are some examples:

- Define clear goals to have a sense of purpose and meaning.

- Exclude distractions and conflicting mental priorities.

- Focus your attention on the problem to avoid self-judgment or fear of failure.

The bottom line is that the mind should be free and focused for a while for creative thinking. As long as the flow lasts, our minds will perform optimally and inventively.

Let Yourself Have Insights

In the context of creative thinking, insight is an idea that occurs to you as if it comes out of nowhere. When an idea pops up in your mind when taking a shower or strolling in the park, when you have your "Aha!" or "Eureka!" moment, that is when you have an insight!

From the ancient Greek scholar Archimedes discovering how to measure the volume of irregular objects in the bathtub to

August Kekule's vision of the chemical structure of Benzene while dreaming of a snake seizing its own tail, and countless other historical examples, flashes of insight usually follow the same basic pattern with these steps:

1. Working hard to solve the problem.

2. Getting stuck, taking a break, stop thinking about the issue.

3. A flash of insight occurring out of the blue, without any conscious thought or effort at the moment.

Although it might look and feel like the mind is doing nothing before the moment of insight, scientific research has shown the opposite. Brain scans show that the brain is working even harder before the moments of insight. People who solve problems with insight generate different brain waves than those who used analytical thinking.

Is it possible to have moments of insight more frequently? Maybe, at least we can try! The process of having flashes of insight is more extensive than the three stages explained above. A good approach to stimulate having insights was presented by James Webb Young in his 1965 book *A Technique for Producing Ideas*. Webb explained the process he followed to have ideas for advertisements. But the process is equally applicable to creative idea generation.

Webb distilled his process into the following five steps:

1. *Gathering knowledge*: This includes both general and topic-specific knowledge.

2. *Thinking deep and hard about the problem*: Young believed in working up to the point of giving up over sheer exhaustion. This is similar to going into the creative flow, as explained by Mihaly Csikszentmihalyi. One can visualize the data gathered in the previous step, use various brainstorming tools to generate and record ideas, combine and refine them, look for patterns, and do anything they can to find workable solutions.

3. *Incubation*: This is when the unconscious mind does its magic. When you feel stuck with a problem, it is better to take a break instead of pushing through the problem with thinking harder. Go for a walk and observe new things. Remember that some downtime is needed for your ideas to ferment and grow inside your head. If you feel you've reached the end of your rope, stop what you're doing. Go out, take a break, or just do something Young proposed to do anything that could stimulate imagination and emotions and make you feel better, such as going to the museum, doing charity work, going to the movies, or reading fiction.

    Keep in mind that while incubation often includes doing nothing, it defers from procrastination. This is because incubation happens after hard work when we have worked ourselves to a standstill. But procrastination happens before hard work.

4. *The Eureka moment*: This is the moment of having an insight, seemingly out of the blue. In reality, this is when our unconscious mind has finished thinking and has come with its winner idea.

5. *Refining the idea*: This is similar to the convergent phase of the thinking process. Even a flash of insight needs critique, expansion, and modification to become an actionable solution.

Expand Your Horizons

Enlarging your mental box and expanding your horizons enhance the chances of developing creative ideas. For example, consuming content that's way outside your background, getting inspired by meeting creative people or visiting interesting places, learning new skills, and hanging out with people from different backgrounds are all stimulating experiences. They can give you the chance to see things from a different perspective and trigger inventiveness in your mind.

You don't need to be necessarily original to be creative. However, it would greatly help if you are keen on observing, listening, learning, and leveraging past experiences and what Mother Nature teaches us. Nature is the most incredible invention machine in our world. Biomimicry, i.e., mimicking the strategies found in the natural world to solve human challenges, has always been a significant source of inspiration.

Many of the products we use today were inspired by nature, from Velcro tape to the high-speed bullet train and sonar to efficient wind turbine blades, all with their fantastic development stories. For example, the Japanese engineers developed the unique nose of Shinkansen high-speed trains after the long narrow beak of water kingfisher bird. In the 1990s, when a Shinkansen train exited a typical train tunnel, it caused a loud booming sound, disturbing the wildlife, the people living nearby, and the train passengers. This was because the train traveling at more than 300 km per hour would compress the air inside the tunnel and caused the booming sound once the compressed air was pushed out.

One of the engineers in the train development team was a birdwatcher. He had witnessed how kingfisher birds dive through the air into the water to catch their prey, and they made very little splash. So, he proposed to mimic the front of the bullet train after the kingfisher beak. Sure enough, they tried the idea, and it not only eliminated the boom but made the train more energy-efficient.

The Best of Both Worlds

Is there a conflict between individual and group brainstorming? Is one superior to the other? When should you think up creative ideas independently and when to engage in collaborative ideation? Does solitude enhance creativity, or is collaborative thinking the way to go? Is it possible to let the ideas have sex without the interactions typically happening in collaborations?

There are a lot of conflicting opinions on whether solitude or collaboration is the best route to creativity. Let's first review some real-world examples and some expert opinions from each side of the spectrum. Then we'll get back to the above questions and try to reach some conclusions in the context of brainstorming.

Solitude has always been associated with creativity. Think of all the great artists and scientists developing groundbreaking works of art and science in their solitude. Some of the most prominent innovators of our time developed their inventions on their own. For example, Sir James Dyson spent four years working alone, making 5000 prototypes before introducing the first Dyson bag-free vacuum cleaner in 2003.

Likewise, solitude allows creative minds to examine many different ideas in isolation without being disturbed or judged. Perhaps no one describes this better than Isaac Asimov, the American science-fiction writer, and professor of biochemistry at Boston University, when he wrote in a 1959 essay on creativity that, *"My feeling is that as far as creativity is concerned, isolation is required. The creative person is, in any case, continually working at it. His mind is shuffling his information at all times, even when he is not conscious of it. (The famous example of Kekule working out the structure of benzene in his sleep is well-known.). The presence of others only inhibits this process since creation is embarrassing. For every new good idea you have, there are a hundred, ten thousand foolish ones, which you naturally do not care to display."*

Steve Wozniak designed Apple I, the first Apple computer, all by himself and introduced it in 1976. He had a simple partnership dynamic with Steve Jobs; that Jobs took care of the marketing, and he invented and constructed the computers. Wozniak and his uninterrupted solitary work sessions were the key factors in the initial success of Apple computers.

Wozniak famously said in his autobiography that, *"I don't believe anything really revolutionary has ever been invented by a committee... I'm going to give you some advice that might be hard to take. That advice is: Work alone... not on the committee. Not on a team."* Who can argue about creativity and productivity with one of the most prominent entrepreneurs of the past century?

On the other side of the spectrum, we can think of Walt Disney, his team spirit, and his emphasis on the importance of the collaborative effort to develop some of the most memorable animations of all time. He often attributed his success to team effort and all the ideas coming from his talented individual. He is quoted to say, *"Of all the things I've done, the most vital is coordinating those who work with me and aiming their efforts at a certain goal"* and *"The whole thing here is the organization. Whatever we accomplish belongs to our entire group, a tribute to our combined effort."*

Disney's approach looks like the antithesis of Steve Wozniak's proceeding quote. So, who got it right?! When it comes to creativity and innovation, is it better to think and work alone, uninterrupted and focused, or collaborate, communicate, and interact as a team? Let's see if creativity experts can help us resolve the conundrum.

Once more, there are conflicting ideas. On the one hand, supporters of collaborative work such as Dr. Keith Sawyer, believe that group thinking and collaboration are crucial for creativity and innovation. But, on the other hand, proponents of independent thinking, such as Susan Cain, the author of *Quiet: The Power of Introverts in a World That Can't Stop Talking*, like to be left alone.

In his book, *Group Genius: The Creative Power of Collaboration*, Keith Sawyer gives many examples of creative innovations resulting from the collaboration. One of these examples is how Orville and Wilbur Wright, two bicycle mechanics from Dayton (Ohio), beat leading scientists to build the first airplane through their relentless and unabated collaboration. Sawyer says:

"The Wrights didn't experience a single moment of insight; rather, their collaboration resulted in a string of successive ideas, each sparking the next."

Wright brothers lived together, and their collaboration was visible. But how about modern airplanes? Hundreds of scientists, engineers, and technicians working across the globe collaborate in an invisible network to design and build modern aircrafts. But, unlike the Wright brothers, their collaboration is not always visible. Sawyer discusses this "invisible collaboration" in detail and provides examples such as the development of mountain bikes to show how a largely invisible long-term collaboration resulted in a new product. Various parts of the mountain bike as we know it today, such as the gear-shifter, the handlebar, the frame, the brakes, were designed in the 1970s and 1980s by different riding enthusiasts in Colorado and California. After that, others still optimized the manufacturing process, created marketing strategies, and gradually modified the bike to appeal to the general population. By 1996, mountain biking was an Olympic sport.

Sawyer explains how even invisible collaboration can enhance creativity:

"Collaboration drives creativity because innovation always emerges from a series of sparks – never a single flash of insight. The Wright brothers had lots of small ideas, each critical to the success of the first powered flight. The mountain bike wasn't commercially viable until many distinct ideas came together. These two stories show how the genius of the group emerges through the sanding and polishing of raw innovation."

Pablo Picasso is quoted to say, "Without great solitude, no serious work is possible." That is what Susan Cain, her proponents, and many introverts believe; that impactful and creative work requires deep, focused, and solitary thinking and practice. Cain refers to social loafing, production blocking, and evaluation apprehension as the main reasons for the failure of group brainstorming. We discussed these three common pitfalls in Chapter 5. But we also reviewed techniques and strategies such as brainwriting and questionstorming to tackle them.

It is fair to say that the opponents of group brainstorming usually refer to traditional brainstorming and ignore the newer techniques discussed in Chapter 4. Psychologists who study creativity believe that it requires both solitude and collaboration. This was the case even for great Steve Wozniak. He got the idea of Apple I while exchanging ideas with the Homebrew Computer Club members, a computer hobbyist group in Menlo Park, California. Wozniak liked to give away the Apple I blueprints so that the other members of the Homebrew Computer Club could build their own computers. It was Steve Jobs who convinced Wozniak to start Apple Computer to produce and sell Apple I. Furthermore, the Macintosh computer, which had a much more innovative graphic user interface, resulted from the collaboration between Steve Jobs and Xerox PARC, the lab which first demonstrated the windows-and-mouse technology. Wozniak indeed made the first Apple computer on his own, but Apple Inc. as we know it today came about only because of multiple collaborations between Jobs, Wozniak, Xerox PARC, and many others.

Do you remember what Isaac Asimov said about thinking in solitude? Even he agreed that isolated thinking is not enough. In the same 1959 essay, he says that *"A meeting of creative people may be desirable for reasons other than the act of creation itself. It seems to me then that the purpose of cerebration sessions is not to think up new ideas but to educate the participants in facts and fact-combinations, in theories and vagrant thoughts."* So, there should be a middle ground with an optimum combination of solo and group thinking.

Let's revisit the questions at the beginning of this section. Is there a conflict between individual and group brainstorming? Is one superior to the other? The best answer to these questions is "It depends!" Solo thinking is more effective in creating ideas, and collaborative thinking is more powerful in combining and refining them. So, individual brainstorming could always act as the starting point for creative idea generation. All ideas originate from someone's "head space." In most cases, thinking and working alone are way more effective in having new ideas. But those same ideas could be merged, reshaped, and improved by the collective thinking power.

In addition to the ideation stage, the type and complexity of the problem are also crucial. The groups usually outperform solo workers when the issue is complex and requiring a wide range of skill sets, which is the case in most professional contexts.

Solving complicated technological, business, or social problems is not possible even if an army of geniuses works without interacting with each other.

Individual and group brainstorming have their utility and added value without conflicting with one another. Asking which one is superior is focusing on the wrong question. Instead, one should ask how to use each approach in the right phase of the process and the correct type of question.

Group brainstorming will stay an essential tool in work environments. However, it is solitude and individual brainstorming that needs more attention from companies and government institutes. The ideal situation would be to use both individual and group brainstorming, the former to create ideas and the latter to improve them and create even more. These two could be combined and repeated in cycles, allowing the team members to mull over the group discussions and feedbacks before joining the next group session.

For combinations of individual and group brainstorming to work, it is essential to have a smooth flow of ideas between the individual and group sessions. For example, if you came up with some stellar ideas during your solo thinking, it is better to present them to the group in written or visual formats. You can also try to predict likely objections to your ideas and find solutions for them. Furthermore, you can prepare a few variants of the concept to represent various possible scenarios.

Some of the group brainstorming techniques discussed in Chapter 4, such as brainwriting, and all online and virtual brainstorming techniques, allow for a combination of individual and group brainstorming. With some adaptations to your specific circumstances, it should be possible to leverage the best of your solo thinking time and the group brainstorming session to get amazingly creative ideas.

The end... almost!

Reviews are not easy to come by. As an independent author with a tiny marketing budget, I rely on readers like you to leave a short review on Amazon. Even if it's just a sentence or two!

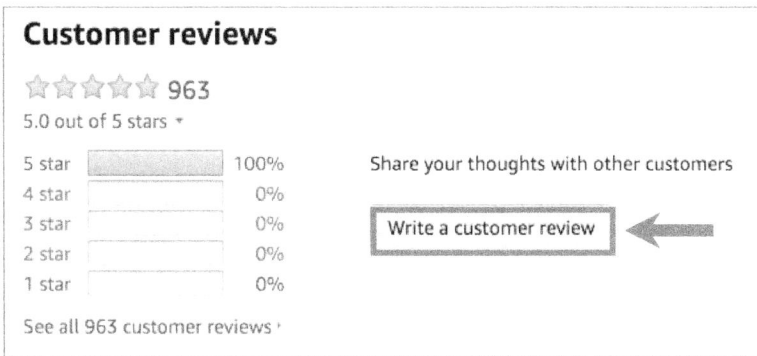

So if you enjoyed the book, please scan this QR code and share your thought!

I am very appreciative for your review as it truly makes a difference. Many thanks for purchasing this book and reading it to the end!

CLOSING REMARKS

Congratulations on finishing the book (almost!)! I hope you've enjoyed it so far and got some valuable new insights on creative thinking and brainstorming. It is time to review the key messages of each chapter and the major takeaways of the book.

We started our journey by asking a core question: How can you generate novel ideas for creative problem-solving? While the outcome of a creative thinking process is unpredictable, the path to those outcomes is illuminated by many lamp posts. This book attempted to identify those lamp posts to guide you towards creative solutions for your small and big problems.

Chapter 1 discussed creativity, its ingredients, thinking techniques, and whether creativity is a universal trait or specific to certain domains. Creative thinking requires four main elements of broad knowledge, purposeful imagination, the mindset of passion, motivation, and persistence, and habitat friendly to creative thinking. Once these ingredients are present, thinking techniques such as productive thinking and divergent-convergent thinking cycles will act like creativity fertilizers.

Finally, collaboration catapults the process by leveraging the synergy effect of collective thinking. The relevance of each ingredient and technique to brainstorming became more apparent as we progressed in the book. For example, the divergent-convergent thinking cycle is the cornerstone of all brainstorming techniques. A key takeaway of Chapter 1 was the domain-specificity of creativity. Domain specificity does not say that a person can only be creative in one domain, only that creativity in one domain is not predictive of creativity in another domain. It also explains why solving complex problems requires collaboration between people with diverse expertise. This is where brainstorming as a well-established collective thinking approach becomes increasingly attractive.

Chapter 2 introduced the history, principles, and practices of traditional brainstorming as formulated and implemented by Alex Osborn in the 1930s-1950s. Osborn proposed the four brainstorming rules of quantity over quality, deferring judgment, encouraging out-of-the-box thinking, and enhancing the ideas by combining and refining them. His meticulous approach and obsessive attention to detail made his brainstorming process such a big success that by the late 1950s, some of America's most prominent corporations and universities utilized brainstorming to generate creative ideas. Since then, brainstorming has spread worldwide and is being used by countless companies and individuals as a standard idea generation tool.

Chapter 3 attempted to characterize the thinking strategies of some of the most creative geniuses of all time, including Albert Einstein, Charles Darwin, and Pablo Picasso. We reviewed strategies such as looking at the problem from different perspectives, gaining inspiration from various sources, building on the existing knowledge, making novel combinations, churning out lots of ideas, and forcing relationships. We also analyzed brainstorming as a design process since the outcome of brainstorming is not known in advance, the same way that artists, poets, writers, designers, or architects do not know exactly how their final product will look like. So, instead, they start with a concept or draft, do iterations, refine the draft, and repeat this cycle until they are satisfied with the outcome. This willingness to tolerate ambiguity and the openness to encountering serendipitous results along the journey are essential ingredients of the right mindset for creative thinking.

Chapter 4 introduced a multitude of brainstorming tools and techniques to approach the problem from different angles, frame the right questions, enhance collaboration between team members, improve engagement, and generate as many great ideas as possible. These techniques address some of the significant shortcomings of traditional brainstorming, including production blocking and evaluation apprehension. This chapter discussed a dozen new variants of conventional brainstorming and the latest virtual and digital brainstorming approaches. You can utilize one or more techniques that best suit your brainstorming needs, hence massively enhancing your odds of having productive and inspiring brainstorming sessions.

Chapters 5 focused on how to have a thriving group brainstorming session, face-to-face or online. We covered effective facilitation, brainstorming mistakes to avoid, and how to leverage the digital tools for better brainstorming. Most of the typical criticisms of brainstorming as an outdated and ineffective approach become irrelevant for a well-prepared and properly facilitated session, which used the correct brainstorming technique. If you plan to organize, facilitate or attend a group brainstorming session, I highly recommend going back to Chapter 5 from time to time as a refresher.

Chapter 6 discussed the value and importance of solitude for creative thinking and tried to debunk the misconception that solo thinking goes against the whole premise of brainstorming as a collective thinking process. The book's final chapter discussed multiple individual brainstorming techniques, strategies to stimulate solo thinking, addressed the conflicting opinions on whether group or individual brainstorming is superior and provided many ideas on how to have the best of both worlds by combining individual and group brainstorming.

Brainstorming has passed the test of time, and we can still learn from it. The keys are to keep the principles in mind and adapt the process for your specific situation. Hopefully, this book will help you in your journey to mastering the art of brainstorming. Who knows how far you can stretch the boundaries of your creative thinking! Thank you for choosing this book, and good luck!

REFERENCES

The Creativity Post | Why We Keep Getting the Same Old Ideas. (2020, November 11). The Creativity Post. https://www.creativitypost.com/create/why_we_keep_getting_t he_same_old_ideas

The Creativity Post | Can Anyone Be Creative? How? What next? (2019, March 29). The Creativity Post. https://www.creativitypost.com/psychology/can_anyone_ be_creative_how_what_next

What is brainstorming and why is it important? | MindManager Blog. (2020, November 16). Mind Manager. https://blog.mindmanager.com/blog/2019/11/what-is-brainstorming-and-why-is-it-important/

Brainstorming and Creative Problem Solving. (2012, April 6). The Alexander W. White Consultancy. https://alexanderwwhite.wordpress.com/designer/type-design/brainstorming-and-creative-problem-solving/

What is Brainstorming And How Is It Helpful? (n.d.). IMindQ. https://www.imindq.com/uses/brainstorming

Mortar, B. (2017, January 18). *How Brainstorming Improves Creative Problem Solving.* Brand & Mortar. https://brandandmortar.com/2017/01/18/how-brainstorming-improves-creative-problem-solving/

J. P. Guilford - New World Encyclopedia. (2021). New World Encyclopedia. https://www.newworldencyclopedia.org/entry/J._P._Guilford

Baer, J. (2015). The Importance of Domain-Specific Expertise in Creativity. Roeper Review, 37(3), 165–178. https://doi.org/10.1080/02783193.2015.1047480

Weinstein, G. (2017). Einstein's Pathway to the Special Theory of Relativity (2nd Edition) (2nd ed.). Cambridge Scholars Publishing.

Michalko, M. (2006). Thinkertoys: A Handbook of Creative-Thinking Techniques. Ten Speed Press.

Besant, H. (2016). The Journey of Brainstorming. *Journal of Transformational Innovation, 2*(1), 1–7.

Rudy, L. J. (2021, July 20). *Starbursting: How to Use Brainstorming Questions to Evaluate Ideas.* Business Envato Tuts. https://business.tutsplus.com/tutorials/starbursting-how-to-use-brainstorming-questions-to-evaluate-ideas--cms-26952

Rudy, L. J. (2021b, July 21). *How to Use Brainwriting for Rapid Idea Generation.* Business Envato Tuts+. https://business.tutsplus.com/tutorials/how-to-use-brainwriting-for-rapid-idea-generation--cms-26451

Rudy, L. J. (2021a, July 14). *How to Use Reverse Brainstorming to Develop Innovative Ideas.* Business Envato Tuts+. https://business.tutsplus.com/tutorials/how-to-use-reverse-brainstorming-to-develop-innovative-ideas--cms-27531.

Rudy, L. J. (2021b, July 19). *What Is Rolestorming? A Useful (+Playful) Group Brainstorming Method.* Business Envato Tuts+. https://business.tutsplus.com/tutorials/what-is-rolestorming-group-brainstorming-method--cms-27245.

Rudy, L. J. (2021b, July 15). *Brain-Netting: How to Brainstorm Online Better in a Distributed Team.* Business Envato Tuts+. https://business.tutsplus.com/tutorials/brain-netting-how-to-brainstorm-online--cms-27387.

De Ruijter, R. (2017, June 16). *The Braindump.* Brightstorming. https://brightstorming.com/the-braindump/?lang=en

Coyne, K. P., & Coyne, S. T. (2011). *Brainsteering: A Better Approach to Breakthrough Ideas* (1st ed.). Harper Business.

Coyne, K. P., & Coyne, S. T. (2021, April 16). *Seven steps to better brainstorming.* McKinsey & Company. https://www.mckinsey.com/business-functions/strategy-and-corporate-finance/our-insights/seven-steps-to-better-brainstorming#

Cole, S. (2014, August 26). *7 Things You Should Never Say When Brainstorming.* Fast Company. https://www.fastcompany.com/3034781/7-things-you-should-never-say-when-brainstorming

Sri, C. (2020, July 29). 5 Brainstorming Questions to Inspire Creativity | MiroBlog. MiroBlog | A Blog by Miro. https://miro.com/blog/brainstorming-questions/

Ogbodo, I. (2020, May 19). *Why I Hate Brainstorming Sessions. - Isaac Ogbodo.* Medium. https://medium.com/@ogbodoisaac/why-i-hate-brainstorming-sessions-b896b7d9a525

Snow, S. (2020, February 13). *The Trouble With Brainstorming (And How To Overcome It).* Contently. https://contently.com/2020/02/06/overcome-brainstorming-problems/

Why Brainstorming Works Better Online. (2015, April 2). Harvard Business Review. https://hbr.org/2015/04/why-brainstorming-works-better-online

Frey, C. (2016, January 20). *Leverage David Bowie's favorite visual brainstorming technique.* Mind Mapping Software Blog. https://mindmappingsoftwareblog.com/david-bowies-favorite-visual-brainstorming-technique/

Salvador Dali's Creative Thinking Technique. (2020, November 9). The Creativity Post. https://www.creativitypost.com/article/salvador_dalis_creative_thinking_technique

Wedell-Wedellsborg, T. (2020). *What's Your Problem?: To Solve Your Toughest Problems, Change the Problems You Solve.* Harvard Business Review Press.

Sawyer, K. (2017). Group Genius: The Creative Power of Collaboration (2nd ed.). Basic Books.

Creative Thinking: The Four Most Powerful Creative Thinking Techniques. (2019, March 28). Mark McGuinness | Creative Coach. https://lateralaction.com/creative-thinking/

Creative Thinking: The Four Most Powerful Creative Thinking Techniques. (2019b, March 28). Mark McGuinness | Creative Coach. https://lateralaction.com/creative-thinking/

Young, J. W. (2019). *A Technique for Producing Ideas.* Independently published.

McGuinness, M. (2019, March 4). *What's the Difference Between Incubation and Procrastination?* Mark McGuinness | Creative Coach, https://www.wishfulthinking.co.uk/2007/10/23/whats-the-difference-between-incubation-and-procrastination/

Schenck, E. (2006). The Houdini Solution: Put Creativity and Innovation to work by thinking inside the box (1st ed.). McGraw-Hill Education.

The Houdini Solution. (2006, December 2). Communication Arts. https://www.commarts.com/columns/the-houdini-solution

Juarez, I. (2018, January 10). *Leave Me Alone and Let Me Innovate, Isabel Juarez.* Medium. https://medium.com/@ijuarez001/leave-me-alone-and-let-me-innovate-edfe22a044cc

Cain, S. (2013). Quiet: The Power of Introverts in a World That Can't Stop Talking, Crown.

Juarez, I. (2018, January 10). *Leave Me Alone and Let Me Innovate - Isabel Juarez*. Medium. https://medium.com/@ijuarez001/leave-me-alone-and-let-me-innovate-edfe22a044cc

K. (2012, January 16). *Does Solitude Enhance Creativity? A Critique of Susan Cain's Attack on Collaboration*. The Creativity Guru. https://keithsawyer.wordpress.com/2012/01/16/does-solitude-enhance-creativity-a-critique-of-susan-cains-attack-on-collaboration/

Torrance Test - an overview | ScienceDirect Topics. (n.d.). Retrieved from https://www.sciencedirect.com/topics/psychology/torrance-test

Wellner, A. S. (2020, February 6). A Perfect Brainstorm. Retrieved from https://www.inc.com/magazine/20031001/strategies.html

The Creativity Post | Allow Your Ideas to Have Sex with Other Ideas. . .. (2019, March 29). Retrieved from https://www.creativitypost.com/psychology/allow_your_ideas_to_have_sex_with_other_ideas_to_create_new_ideas

J. (2013, November 29). 8 methods that anyone can use to think like a genius | Game-Changer. Retrieved from http://www.game-changer.net/2009/05/21/how-to-think-like-a-genius/#.X6plmYhKhPY

Productive thinking vs Reproductive thinking. (n.d.). Retrieved from:

http://sdoulger.blogspot.com/2011/01/productive-thinking-vs-reproductive.html

Farmers Insurance. (n.d.). How This Process Will Totally Transform Your Brainstorming Sessions. Retrieved from https://www.inc.com/farmers-insurance/forget-everything-you-know-about-brainstorming-do-this-instead-to-generate-ideas.html?cid=search

Bhasin, H. (2020, February 28). https://www.marketing91.com/concept-development/. Retrieved from https://www.marketing91.com/concept-developmcnt/

Torabi, N. (2020, September 6). *'Design Thinking' — steering the 'Inspiration phase.'* Medium. https://neemz.medium.com/design-thinking-steering-the-inspiration-phase-36cd53f6feaf

Brainstorming. (n.d.). Tutorials.Istudy.Psu.Edu. http://tutorials.istudy.psu.edu/brainstorming/

Seelig, T. (n.d.). *In Genius: A Crash Course on Creativity.* NA, USA: HarperOne.

Wikipedia contributors. (2021, March 24). *New Coke.* Wikipedia. https://en.wikipedia.org/wiki/New_Coke

Lehrer, J. (2012). *Imagine: How Creativity Works.* NA, USA: Houghton Mifflin Harcourt.

Vaičiulaitytė, G. (2019, March 14). *100 Genius Solutions To Everyday Problems You Didn't Know Existed.* Bored Panda. https://www.boredpanda.com/creative-solutions-everyday-problems/?utm_source=google&utm_medium=organic&utm_campaign=organic

Control, V. (2021, March 19). *8 Great Design Thinking Examples*. Voltage Control. https://voltagecontrol.com/blog/8-great-design-thinking-examples/

Ridley, M. (2011). *The Rational Optimist: How Prosperity Evolves (P.S.)* (Illustrated ed.). Harper Perennial.

Gladwell, M. (2011). *Outliers: The Story of Success* (1st ed.). Back Bay Books.

The Creativity Post | 101 Tips on How to Become More Creative. (2019, March 29). Retrieved from https://www.creativitypost.com/create/101_tips_on_how_to_become_more_creative

Creative thinking - how to get out of the box and generate ideas: Giovanni Corazza at TEDxRoma. (2014, March 11). [Video file]. Retrieved from https://www.youtube.com/watch?v=bEusrD8g-dM

Ware, S. (2021, January 6). Six Examples of Creativity at Work. Retrieved from https://truscribe.com/six-examples-of-creativity-at-work/

Bathla, S., & McCoy, R. (2019). Think out of the Box: Generate Ideas on Demand, Improve Problem Solving, Make Better Decisions, and Start Thinking Your Way to the Top. NA, USA: Som Bathla.

T. (2016, March 25). Conceptual Blending. Retrieved from https://thinkibility.com/2016/03/25/conceptual-blending/

The Importance of Domain-Specific Expertise in Creativity. (n.d.). Retrieved from https://www.tandfonline.com/doi/full/10.1080/02783193.2015.1047480

Blog post: Conceptual Blending for Creative Thinking. (2015, August 25). K-12 Thoughtful Learning. https://k12.thoughtfullearning.com/blogpost/conceptual-blending-creative-thinking

Mansfield, D. (2018, October 1). *15 Creative Exercises That Are Better Than Brainstorming.* Hubspot. https://blog.hubspot.com/marketing/creative-exercises-better-than-brainstorming

Asimov, I. (2020, April 2). *Isaac Asimov Asks, "How Do People Get New Ideas?"* MIT Technology Review. https://www.technologyreview.com/2014/10/20/169899/isaac-asimov-asks-how-do-people-get-new-ideas/